OPTICAL THEORIES

OPTICAL THEORIES

BASED ON LECTURES DELIVERED BEFORE
THE CALCUTTA UNIVERSITY

BY

D. N. MALLIK, B.A. (Cantab.), Sc.D. (Dub.), F.R.S.E.

Late Scholar, Peterhouse, Cambridge
Professor, Presidency College, Calcutta
Fellow of the Calcutta University

SECOND EDITION
(REVISED)

Cambridge:
at the University Press
1921

CAMBRIDGE
UNIVERSITY PRESS

University Printing House, Cambridge CB2 8BS, United Kingdom

Cambridge University Press is part of the University of Cambridge.

It furthers the University's mission by disseminating knowledge in the pursuit of education, learning and research at the highest international levels of excellence.

www.cambridge.org
Information on this title: www.cambridge.org/9781316611838

First edition 1917
Second edition 1921
First paperback edition 2016

A catalogue record for this publication is available from the British Library

ISBN 978-1-316-61183-8 Paperback

PREFACE

IN the year 1912, the University of Calcutta appointed me Reader in Physics and invited me to deliver a course of lectures to its advanced students on Optical Theories, one of the conditions of the appointment being that the lectures should be published after their delivery. The lectures were actually delivered during the months of February and March, 1912, but pressure of other work has prevented me, till now, from seeing them through the press.

It has been my object in these lectures to trace the development of Optical Theories from the earliest times to the present day. I have tried to understand and help others (so far as I can) to understand the relation between the different theories, so that one may be clear as to how much is certainly known and how much is mere speculation. In the midst of the bewildering mass of investigations that a student of luminiferous medium is confronted with at the present day, a sketch, such as the one attempted here, describing, with such details as will make the general argument intelligible, how we have been led up to the present position and what that position really is should, as it seems to me, be of considerable use. How far I have succeeded in my attempt, it is for others to judge.

To the latest developments of the optical theory including the theory of relativity, no reference has been made here. I hope to deal with them in a later volume, if the present attempt proves successful.

D. N. M.

November, 1916.

PREFACE TO THE SECOND EDITION

IN the present edition, I have included a brief account of the theory of relativity and the quantum theory which had been previously left out for a separate and special treatment. It has since appeared to me, however, that no statement of the present position can be regarded as at all satisfactory at the present day, which omits a reference to them altogether.

The subject matter of the present treatise—the nature of the electro-magnetic field—is, in reality, the one great general problem of modern Physics. For "if," to quote Larmor, "it is correct to say with Maxwell that all radiation is an electrodynamic phenomenon, it is equally correct to say with him that electrodynamic relations between material bodies are established by the operation, on the molecules of those bodies, of fields of force, which are propagated in free space, as radiation and in accordance with the laws of radiation from one body to another." And it seemed to be desirable that the points of view which the new theories represent, should be stated, however briefly, in order that the nature of this problem and the attempted solutions may be clearly brought into view.

D. N. MALLIK.

March, 1920.

CONTENTS

CHAP.		PAGE
I.	EARLY SPECULATIONS. CORPUSCULAR THEORY, UNDULATORY THEORY	1
II.	ELASTIC SOLID THEORY	33
III.	ELECTRO-MAGNETIC THEORY	61
IV.	ELECTRON THEORY	124
V.	THEORY OF RELATIVITY	163
VI.	SUMMARY. CONCLUSIONS	176
	APPENDIX I	187
	II	190
	III	191
	IV	192
	V	196
	VI	198
	INDEX	201

CHAPTER I

INTRODUCTION. CORPUSCULAR THEORY. UNDULATORY THEORY

SECTION I. ANCIENT SPECULATIONS. DESCARTES. FERMAT

1. A complete theory of optics has to furnish an adequate account, not merely of the nature of light but also of the mode and mechanism of its propagation, as well as the nature of the medium in which the propagation takes place. And we shall see, as we pass in review, in historical order, the various theories that have been proposed, that our knowledge on these points is after all extremely limited.

2. This, indeed, is what is to be expected—illustrating, we may remark in passing, the limit and scope of scientific inquiry in general. For, although the phenomena with which, to confine ourselves to optical investigations alone, we have to deal are simple and well-known, and although we can formulate the laws governing them—like any other body of scientific laws—which state (as Karl Pearson has put it), "in conceptual shorthand, that routine of our perceptions, which forms for us the totality of the phenomena," to which they refer, when we come to inquire into the intimate nature of the processes, associated with these laws, we are confronted with insuperable difficulties. These are so subtle and deep-seated, that they will probably always elude our grasp. We are, therefore, reduced to preparing models, that shall approximate to the actual, as far as possible, and our task consists really in improving these models, more and more, so that they may more and more nearly approximate to the actual. Thus, of the mode of propagation of light and the nature of the medium which takes part in its propagation, we, with our limitations, can never have any direct knowledge.

We are, therefore, driven to reason by analogy and construct a picture which shall be as near a representation of the actual as we can make it. At first, only the outlines are noted; then, the details are gradually filled in, but the picture can never fully represent the actual, unless and until our powers of perception are so sharpened that we are able to take cognizance of the order of quantities which we now call infinitesimal.

3. Although speculations as to the nature of light are of remote antiquity, anything like a clear notion as to the nature of light belongs only to comparatively modern times.

4. Among the most ancient recorded speculations on the subject of light seem to have been those of the Hindus. In the *Nyaya Vasya**, there is a discussion, from the point of view of the Nyaya Philosophy, on the theory of the mirage:

"In the summer, the rays of the sun coming in contact with the heat issuing from the earth vibrate upwards and downwards and gradually reaching the distant observer's eye produce a false impression of water, by an incorrect association of sight and object."

It is clear that the writer of *Nyaya Vasya* was interested only in the metaphysics of the subject. For, with regard to this discussion, the commentary *Bartika* points out that in this case, although the solar rays are certainly there, as well as the vibrations, what is at fault is the impression of water—*the mistaken association of the phenomenon with water*, etc. All the same, it is evident from this, that the above was the generally accepted theory of the mirage at the time the *Vasya* was written and must have been treated of in formal books on Physics, which seem to have been lost.

* The Indian schools of Philosophy (called *Darsanas*) are classed under two main heads. One class, called the '*Astika*' (the believing), believes in the authority of the *Vedas*, and the other called the '*Nastika*' (non-believing) does not recognize this authority. Each of these is divided into six schools: *Nyaya*, *Vaisesika*, *Sankhya*, *Patanjal*, *Purva Mimansa* and *Uttara Mimansa* (or *Vedanta*) belong to the first or the Astika group, while *Charvak*, the Jain and four Buddhistic schools belong to the latter. *Nyaya Sutra*, of which the reputed author was *Gautoma*, the founder of the Nyaya school, deals with canons of correct reasoning. *Nyaya Vasya* is a commentary (by *Vatsayana*) on *Nyaya Sutra* and *Bartika* (by *Udyotaka*) is a commentary on the latter.

5. In the same work (*Nyaya Vasya*), the following explanation occurs of the formation of images by reflection:

"The 'eye rays,' striking against mirrors, return and come in contact with the face [to which the eye belongs, i.e., of the observer]. From this contact is derived the knowledge of the face. The '*rupa*,' i.e., form or colour of the mirror, helps in producing this knowledge."

The commentary *Bartika* thus amplifies the theory:

"The 'eye rays' rebound at mirrors, water, etc. On rebounding, they come in contact with the face. As the extremities (fore-part) of rays come into relation with the face, the face appears to be in front. This is the law relating to knowledge acquired by means of the eye, viz., the object which comes into relation with the extremities (fore-part) of the 'eye rays' is made out by this knowledge to be *in front*, e.g., the knowledge of the face of a man standing in front [of the observer]."

6. Before the days of *Nyaya Sutra*, a theory prevailed that every object emits rays*. The author of *Nyaya Sutra* points out that in that case, stones, etc. should be capable of being seen at night, and the author of *Nyaya Vasya* further argues that one could not imagine such a thing (viz., rays emitted by stones); whereas 'eye rays' are imaginable!

7 On the theory of transparency, we are told, in *Nyaya Sutra*, that "the 'eye rays' are not turned back by (i.e., are allowed to pass through) glass, etc. This is how objects placed beyond glass, etc. can come into relation with 'eye rays' and are seen. Opaque bodies like walls turn off 'eye rays' and therefore bodies cannot be seen through them."

8. As to the eye rays, it is stated in *Nyaya Kandali*, a treatise of the *Vaisesika* school, that "their form cannot be seen nor can they be touched, but they go to a distance and produce the knowledge of bodies, if nothing stands in the way." Moreover, the eye rays, like solar rays, are to be regarded (according to a commentary on the *Vedanta Parivasa* of the *Vedanta* school) as "*transparent bodies* [Art. 10] and may therefore *have rapid motion*."

* Cf. Pythagoras [Art. 9], who is said to have received his early training in India.

1—2

9. With regard to these extracts*, it should be noted that they only incidentally occur as illustrations of principles discussed in Hindu Philosophy. They have, therefore, no further interest from our present point of view, beyond the fact that they show that optical speculations in India dated beyond the days of the *Nyaya Sutra.* It is interesting, moreover, to note that similar speculations appear in the first systematic European work on light, that of Empedocles (444 B.C.). According to him, light consists of particles, projected from luminous bodies, and a vision is the effect of these bodies and a *visual influence*, emitted by the eye itself, although Pythagoras and his followers had previously maintained that vision was caused by particles continually projected from the surfaces of objects into the pupil of the eye. If, therefore, as is maintained by some, *Nyaya Sutra* was written between 500 B.C. and 200 B.C., it would follow that contemporary optical ideas in Greece and India proceeded on similar lines.

10. The fallacy of the theory of a 'visual influence' was discussed by Aristotle (350 B.C.), who argued that if a visual influence was emitted by the eye, we should be able to see in the dark. He considered it more probable that light consisted in an impulse, propagated through a continuous medium, rather than an emanation of distinct particles. Light, according to him, is the action of a transparent substance [Art. 8] and if there were absolutely no medium between the eye and any visible object, it would be absolutely impossible that we should see it. The meaning of the latter part of the argument seems to be that if, between the luminous object and the eye receiving the impression, there did not exist something endued with the physical property that makes it capable of transmitting the influence (whatever its nature may be) emitted by the luminous object, that influence could never reach the eye. This is, also, in effect the postulate of modern science.

11. From this time up to that of Descartes, optical discoveries related mainly to the two fundamental phenomena

* I am indebted to Mahamahopadhyaya Gurucharan Tarkadarsana-tirtha, Professor of Nyaya, Sanskrit College, Calcutta, for these extracts.

of reflection and refraction. Archimedes was evidently acquainted with the property of burning mirrors and seems to have made some experimental investigations on this subject, while Vitellio, a Pole, developed a mathematical theory of optics*. Roger Bacon is said to have invented the magic lantern, and is regarded by some as the first to have invented the telescope also. But the first person who is certainly known to have made a telescope was Janson, a Dutchman, whose son, by accidentally placing a concave and a convex spectacle glass at a short distance from each other, observed the increased apparent magnitude of an object seen through them. It was to Galileo, however, that we owe the first construction and use of such a telescope (the Galilean) for astronomical observations. And it was to him also that we owe its theory. Galileo states†, in his *Nuncius Sidereus*, that happening to hear that a Belgian had invented a perspective instrument by means of which distant objects appeared nearer and larger, he discovered its construction by considering the effects of refraction. Finally, Kepler worked out the true theory of the Astronomical telescope (a combination of convex lenses), made some experiments on the nature of coloured bodies and experimentally verified the formation of inverted images on the retina of the eye.

12. Descartes published the law of refraction, originally discovered by Snell, and deduced the law from theory: or rather an analogy. "When rays meet ponderable bodies, they are liable to be deflected or stopped in the same way as the motion of a ball or stone impinging on another body."

Let a ball‡ thrown from A meet at B a cloth CBE, so weak that the ball is able to break through it and pass beyond, but with its resultant velocity reduced in some definite proportion, say $1 : k$.

Then, if BI = length, measured on the refracted ray = AB, the time to describe $BI = k \times$ the time to describe AB.

But the component velocity parallel to the cloth is unaffected.

* Lectures by Th. Young. † *Ency. Brit.*

‡ Whittaker, *A History of Theories of the Ether and Electricity.*

$$\therefore\ BE = \text{projection of } BI, \text{ on the cloth,}$$
$$= k\,.\,BC, \text{ where } BC \text{ is the projection of } AB.$$
$$\therefore \text{ if } i = \angle CAB,$$
$$r = \angle BIE,$$
$$\sin r = \frac{BE}{BI} = k\,.\,\frac{CB}{BA} = k\,.\,\sin i,$$

or the sines of the angles of incidence and refraction are in a constant ratio*.

13. He also propounded a theory of light. On this theory, light consists in pressure transmitted instantaneously through a medium, infinitely elastic, while colour, according to him, is due to a rotatory motion of the particles of the medium, the particles which rotate most rapidly giving the sensation of red, etc. But Descartes supposed [Art. 12] light to pass more quickly through a denser than a rarer medium, while Fermat, maintaining the contrary view, enunciated the principle of swiftest propagation of light.

14. Fermat's argument was metaphysical—"nature works by the shortest route." The result, however, is remarkably correct.

For, this law [Fermat's law of swiftest propagation of light] states that $\delta\int dt = 0$, where t is the time of propagation of light between two given points and δ is the operator of the calculus of variation. Now, since, on the wave theory, μ, the index of refraction, varies inversely as the velocity of propagation, the above is obviously the same as $\delta\int \mu ds = 0$, an equation which, as we shall see, analytically embodies a complete kinematical statement of all optical phenomena.

15. Thus, $$\delta\int \mu ds = \int \delta\mu ds + \int \mu d\delta s$$

$$= \int\left(\frac{\partial\mu}{\partial x}\delta x + \frac{\partial\mu}{\partial y}\delta y + \frac{\partial\mu}{\partial z}\delta z\right) ds + \int \mu\left(\frac{dx}{ds}\, d\delta x + \dots + \dots\right)$$

since $$ds^2 = dx^2 + dy^2 + dz^2$$

and therefore $$d\delta s = \left(\frac{dx}{ds}\, d\delta x + \dots + \dots\right).$$

* If v, v' be the velocities in the two media, $v'/v = 1/k$, and $v' \sin r = v \sin i$.

Therefore, integrating the second term by parts, we have

$$\left[\mu\frac{dx}{ds}\right]_1^2 \delta x + \ldots + \int\left[\frac{\partial\mu}{\partial x} - \frac{d}{ds}\left(\mu\frac{dx}{ds}\right)\right]\delta x ds + \ldots + \ldots = 0$$

or, since δx, δy, δz are independent,

$$\frac{\partial\mu}{\partial x} = \frac{d}{ds}\left(\mu\frac{dx}{ds}\right), \text{ etc.,}$$

while, at a surface bounding media (1, 2), $\mu\frac{dx}{ds}$, etc., are continuous, giving the ordinary laws of reflection and refraction and the path of a ray in a heterogeneous singly-refracting medium.

16. Again, let PQ, QR be the incident and refracted rays at a surface of separation between the media of refractive indices μ, μ' and let $Q''R''$ be a ray parallel to PQ, and PqR a ray consecutive to PQR.

Then, drawing $QQ'Q''$ perpendicular to PQ and qq', RR'' perpendicular to QR, we have

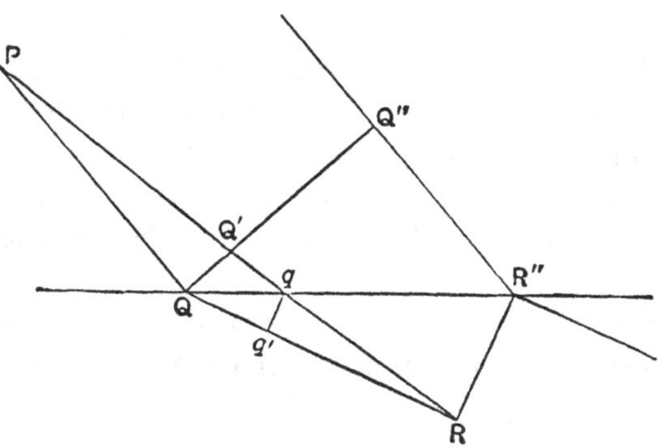

since $$\mu PQ + \mu' QR = \mu Pq + \mu' qR,$$

$$\mu Q'q = \mu' Qq',$$

or $$\mu Q''R'' = \mu' QR.$$

This gives Huyghens' construction for singly-refracting media.

17. If the second medium is doubly-refracting, QQ', RR'' are the traces of the wave fronts (in the plane of the paper) and thus, since the wave surface is the envelope of the wave front, the wave surface can be determined, in the usual way [Art. 54], as the envelope of

$$lx + my + nz = v,$$

where l, m, n are the direction-cosines of the normal to the wave front, and v the velocity of propagation, while l, m, n and v must be connected by a relation which, however, requires to be determined on some theory independent of Fermat's principle (such as Fresnel's). Thus, Fermat's principle is seen to be capable of giving a complete kinematical account of double refraction, also—with the help of a subsidiary hypothesis, regarding the law of variation of μ with direction.

18. Again, since $\delta \int \mu ds = 0$ we may take $\mu ds = dV$, and we have $\delta V = 0$ at each reflection or refraction.

And, since

$$\frac{\partial V}{\partial s} = \mu, \text{ or } \frac{\partial V}{\partial x} : \frac{\partial V}{\partial y} : \frac{\partial V}{\partial z} :: \alpha . \beta . \gamma.$$

where α, β, γ are the direction-cosines of the ray, we conclude that $V =$ constant is a surface orthogonal to a system of rays.

Accordingly, if a given surface can, at any stage, be made to coincide with a surface $V =$ constant, a surface can always be drawn to coincide with any other member of the family $V =$ constant: Or

Any system of rays originally orthogonal to a surface will always be orthogonal to a surface, after any number of reflections and refractions.

19. Since, then, the rays of light (which are orthogonal to a surface) may be regarded as normals to a family of surfaces, and, by Sturm's theorem, all the normals to a surface in the neighbourhood of a point converge to or diverge from two focal lines at right angles to one another, each of which passes through the centre of curvature of one of the principal normal sections and is perpendicular to the plane of that section, we conclude that all the rays of a thin pencil which can be cut at right angles by a surface pass through two lines, such that the planes containing either of them and the principal ray are perpendicular to each other.

20. Again, since the equation of a surface near the origin with the axis of z along the normal at the origin is, to the second order,

$$2z = \frac{x^2}{\rho_1} + \frac{y^2}{\rho_2},$$

the *characteristic function* (V = constant) for a thin pencil in a medium μ, with the axial ray proceeding along the axis of z, is

$$V = \mu\left(2z - \frac{x^2}{\rho_1} - \frac{y^2}{\rho_2}\right),$$

approximately, if aberration is neglected. With the help of this equation, the problem of reflection and refraction of direct and oblique pencils (aberration being neglected) can be treated in the usual way.

21. To illustrate this, consider the following example:

A pencil of rays is refracted through a prism and the axis of the pencil is constantly in a principal plane of the prism: also the angular position of the focal lines at incidence and after the first and second refraction relative to the edge of the prism is defined by α, β, γ. If, now, the distances of the initial and final foci from the first and second surfaces are u_1, u_2 and v_1, v_2, respectively, to obtain the equations of refraction:—

Let us take U_1, U_2 as the distances of the foci after the first refraction from the first surface, and V_1, V_2, those from the second surface.

Then $V_1 - U_1 = V_2 - U_2$ = length of the axial ray in the prism.

Let, finally, ϕ, ϕ' be the angles of incidence and refraction at the first surface, and ψ, ψ', those at the second surface.

The edge of the prism being taken as the axis of y and the normal to the face of incidence as the axis of z, the direction-cosines of one of the focal lines before incidence are

$$\sin\alpha\cos\phi, \quad \cos\alpha, \quad \sin\alpha\sin\phi.$$

Thus, the characteristic function, before incidence, becomes

$$V = \mu \left\{x \sin\phi + z\cos\phi - \frac{1}{2u_1}(x\sin\alpha\cos\phi + y\cos\alpha + z\sin\alpha\sin\phi)^2 - \frac{1}{2u_2}(x\cos\alpha\cos\phi - y\sin\alpha + z\cos\alpha\sin\phi)^2\right\}.$$

From the continuity of the function, at $z = 0$, we get, by equating coefficients of x, x^2, xy, and y^2,

$$\mu\sin\phi = \mu'\sin\phi',$$

$$\mu\left(\frac{\sin^2\alpha}{u_1} + \frac{\cos^2\alpha}{u_2}\right)\cos^2\phi = \mu'\left(\frac{\sin^2\beta}{U_1} + \frac{\cos^2\beta}{U_2}\right),$$

$$\mu\left(\frac{\cos^2\alpha}{u_1} + \frac{\sin^2\alpha}{u_2}\right) = \mu'\left(\frac{\cos^2\beta}{U_1} + \frac{\sin^2\beta}{U_2}\right),$$

$$\mu\sin\alpha\cos\alpha\left(\frac{1}{u_1} - \frac{1}{u_2}\right)\cos\phi = \mu'\sin\beta\cos\beta\left(\frac{1}{U_1} - \frac{1}{U_2}\right)\cos\phi',$$

and similar equations for the second refraction.

22. In order to take account of aberration, we must obtain the characteristic function up to the order zx^2.

Let the equation of the characteristic surface ($V=0$) be

$$2z = \frac{x^2}{\rho_1} + \frac{y^2}{\rho_2} + 2z(\alpha x^2 + \beta xy + \gamma y^2) + \text{etc.}$$

Now the perpendicular from the origin on the tangent plane at x', y', z' to the surface $2z = \frac{x^2}{\rho_1} + \frac{y^2}{\rho_2}$ is

$$\frac{z'}{\sqrt{1 + \frac{x'^2}{\rho_1^2} + \frac{y'^2}{\rho_2^2}}} = z'\left(1 - \frac{1}{2}\frac{x'^2}{\rho_1^2} - \frac{1}{2}\frac{y'^2}{\rho_2^2}\right),$$

if terms of the order zx^2 are retained.

Therefore, in $$2z = \frac{x^2}{\rho_1} + \frac{y^2}{\rho_2}$$

we must write $z\left(1 - \frac{1}{2}\frac{x^2}{\rho_1^2} - \frac{1}{2}\frac{y^2}{\rho_2^2}\right)$ instead of z,

so that the result may be correct up to this order.

Hence the equation of the characteristic surface ($V=$ constant) is of the form

$$V + z = \frac{1}{2}\frac{x^2}{\rho_1} + \frac{1}{2}\frac{y^2}{\rho_2} + z\left(\frac{1}{2}\frac{x^2}{\rho_1^2} + \frac{1}{2}\frac{y^2}{\rho_2^2}\right) + ax^3 + 3bx^2y + 3cxy^2 + dy^3 \text{ (say)},$$

the terms in z^2 and z^3 being neglected, as being of higher orders.

The projection of this on the principal plane $y=0$ is

$$z=\frac{1}{2}\frac{x^2}{\rho_1}+\frac{1}{2}z\frac{x^2}{\rho_1^2}+ax^3.$$

The equation of the normal to this at x', z' is

$$\frac{x-x'}{\frac{x'}{\rho_1}+z\frac{x'}{\rho_1^2}+3ax'^2}=\frac{z-z'}{\frac{1}{2}\frac{x'^2}{\rho_1^2}-1}.$$

This meets the normal through the origin at the point $Z, 0$,

where
$$Z=z'-\frac{\frac{1}{2}\frac{x'^2}{\rho_1^2}-1}{\frac{1}{\rho_1}+\frac{z'}{\rho_1^2}+3ax'}.$$

Therefore, the aberration in the principal plane is given by

$$Z-\rho_1=z'-\frac{\frac{1}{2}\frac{x'^2}{\rho_1^2}-1}{\frac{1}{\rho_1}+\frac{z'}{\rho_1^2}+3ax'}-\rho_1$$

$$=z'-\frac{\left[\frac{1}{2}\frac{x'^2}{\rho_1^2}+\frac{z'}{\rho_1}+3ax'\rho_1\right]}{\frac{1}{\rho_1}+\frac{z'}{\rho_1^2}+3ax'}$$

$$=-3ax'\rho_1^2, \text{ rejecting terms of higher orders,}$$

$$=-3a\rho_1^3\theta_1, \text{ where } \frac{x'}{\rho_1}=\theta_1.$$

We conclude, therefore, that all the results of geometrical optics can be deduced from Fermat's law.

23. Proceeding, now, to the dynamical significance of Fermat's law (which, as we have seen, Fermat deduced from metaphysical considerations), we observe that the configuration of equilibrium and motion of a dynamical system is defined by

$$\delta\int(T-V)\,dt=0,$$

where $T =$ Kinetic energy,

$V =$ Potential energy.

This will be consistent with Fermat's law, if we take, for the propagation of light,

$$T - V = C \text{ (constant)} \quad \text{..................(1).}$$

Also if $V' =$ Potential energy of optical disturbance, during the displacement of the disturbance through the length of a wave,

$$\int T dt = \int V' dt \quad \text{......................(2).}$$

From (1) and (2) we get

$$V + C = V',$$

showing that we must postulate a certain *intrinsic energy* in the system whose motion is associated with the propagation of light, in order to justify the principle of swiftest propagation. [Art. 96.]

24. Again, from the principle of energy, we have

$$T + V = C' \text{ (constant)} \quad \text{..................(3).}$$

Therefore from (1) and (3) we get

$$\left.\begin{aligned} 2T &= C' + C \\ 2V &= C' - C \end{aligned}\right\}.$$

But this is meaningless, since the *mean* potential energy and the mean kinetic energy are alone constant, as these quantities are taken to mean in the above equations. Accordingly, the only conclusion that seems to be consistent with all the equations is that *the optical energy is entirely kinetic.*

The equation (1) then becomes $T = A$, the equation (2) disappears and the equation (3) becomes $T' = A'$, and

$$T' - T = A' - A = \text{intrinsic energy.}$$

Again, if the potential energy of deformation of the ethereal medium involved in the propagation of light is to be regarded as essentially kinetic, we are led to conclude that *all energy is kinetic.* [See Appendix I.]

25. This view of the intimate nature of energy intrinsic or otherwise is partially accepted, in effect, in different branches of Physics. Thus, on the kinetic theory of gases, the pressure of a gas has a kinetic origin, and, in fact, all cases of equilibrium in molecular physics are best explained as those of mobile or convective equilibrium.

26. But, if the interpretation of Fermat's law, sketched above, is admissible, we are led to a further generalization,—to regard the energy of mere configuration, also, as kinetic. We must, in fact, conceive some subtle ethereal and electronic [Ch. IV] motion, being associated with every given configuration of a conservative system, existing in the field; thus the so-called potential energy of a vibrating system at any moment is, in reality, kinetic energy of the field. When we consider the later theories, we shall see that there is considerable justification for the view that the energy of light is entirely kinetic*.

Section II. Hooke. Newton

27. In 1665, Hooke published a theory of light in his *Micrographia*. According to him, light is "a quick and short vibratory motion, propagated in every way through a homogeneous highly elastic medium in straight lines, like rays from the centre of a sphere." Hooke believed that refraction is produced by the readier transmission of light through the denser medium and he gave a geometrical construction, similar to Huyghens', for the incident and refracted wave-fronts. According to Hooke, colour is generated by the distortion of the disturbance in the course of refraction.

28. Newton† was evidently aware of this theory, a portion of which, that relating to colour, he refuted on the result of the first of his optical researches. But it was a *hypothesis* merely, unsupported by facts and experiments and was moreover beset with formidable difficulties. "For, to me," said Newton, "the fundamental supposition itself seems impossible, viz., that waves or vibrations of any fluid can, like the rays of light, be propagated in straight lines without continued and very extravagant spreading and bending into the quiescent medium, where they

* Fermat's Law. *Philosophical Magazine*, July, 1913.

† Glazebrook, Address as President of the Physical Section of the British Association, 1893.

are terminated by it." The corpuscular theory, on the other hand, seemed to him to be capable of offering a dynamical (or a quasi-dynamical) explanation of known optical phenomena, and Newton, accordingly, applied it to the explanation of reflection, refraction, diffraction, colours of thin plates and even polarisation. In doing so, he was forced to subtle postulates, which it would not be without interest to examine briefly.

29. (*a*) Reflection.

Let MN limit the region of molecular activity, AB being the reflecting surface. Let XY be a ray, that is, the path of a light corpuscle, which when it reaches the limit of the region of molecular activity, viz. MN at Y, must be regarded as being in a condition to be repelled.

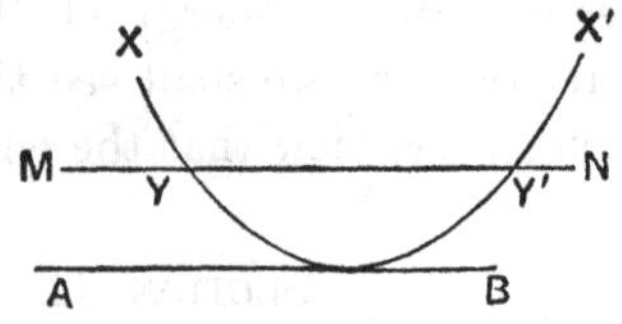

Thus, the corpuscle is acted upon by a repulsive force in a direction perpendicular to AB, so that the normal velocity of the corpuscle decreases. We must assume that it becomes zero when the corpuscle reaches AB, while the tangential velocity remains unchanged.

Hence, ultimately the corpuscle passes off along $Y'X'$ inclined at the same angle to AB as XY. This is the well-known law of reflection.

(*b*) Refraction from a rarer to a denser medium.

The action of the medium on the moving corpuscle is now to be regarded as attractive that is, we must assume that when it enters the region of molecular activity within the denser medium, it is attracted.

The path will, obviously, be bent towards the normal. The velocity in the direction of the normal increases and the tangential velocity remains unchanged.

Thus, let V_a = velocity in air,

V_g = velocity in (say) glass,

i = the angle of incidence,

r = the angle of refraction.

Then, $V_a \sin i = V_g \sin r$:

$$\therefore \quad \frac{\sin i}{\sin r} = \frac{V_g}{V_a} = \mu_g = \text{index of refraction of glass.}$$

(*c*) Refraction from a denser to a rarer medium.

The action of the medium must be regarded as repulsive, so that the velocity perpendicular to the bounding surface must decrease but must not be zero when the corpuscle passes beyond the range of molecular action.

30. We have thus to postulate a repulsion [of suitable magnitude] to explain reflection, an attraction [of suitable magnitude] to explain refraction into a denser medium and a properly adjusted repulsion to explain refraction into a rarer medium*.

31. These postulates are thus explained by Newton (*Optics*, Prop. XII):

"Every ray of light, in its passage through any refracting surface, is put into a certain transient constitution or state, which in the progress of the ray returns at equal intervals and disposes the ray at every return to be easily transmitted through the next refracting surface, and between the returns to be easily reflected by it."

This is manifest (according to Newton) from the arrangement of colours of a thin plate in the reflected and refracted light. For, one and the same sort of rays, at equal angles of incidence on any thin transparent plate, is alternately reflected and transmitted, for many successions, according as the thickness of the plate increases in arithmetical progression:

(1, 3, 5, for reflection; 0, 2, 4, for refraction).

And he proceeds:

"The returns of the disposition of any ray, to be reflected, I will call its Fits of easy reflexion, and those of its disposition to be transmitted, its Fits of easy transmission, and the space it passes, between every return and its next return, the Interval

* It is easy to see that we can explain these as resulting from the motion of electrons suitably interspersed in the ethereal medium.

of its Fits." And, further (Prop. XIII): "The reason why the surfaces of all thick transparent bodies reflect part of the light incident on them and refract the rest, is that some rays at their incidence are in Fits of easy reflexion and others in Fits of easy transmission."

A possible explanation of these fits is supplied in Query 17. "When a ray of light falls upon the surface of any pellucid body and is then refracted and reflected, may not waves of vibrations or tremors be thereby excited in the reflecting medium and do they not overtake the rays of light and, by overtaking them successively, do they not put them into Fits of easy reflexion and easy transmission * ?"

As a further explanation, he suggests (Q. 26) that rays of light may have several sides—four sides to explain double refraction.

32. We have seen that the colours of thin plates were explained by—they, in fact, led to—the hypothesis of fits and a certain polarity of the rays. But, in order to explain all the observed facts, it was necessary to suppose that the length of a fit varies as the secant of the angle of incidence. This has been held by some to emphasize the very artificial nature of the whole theory, but this conclusion is not justified, for it should be noted that it might have conceivably been the consequence of a simpler law †.

His explanation of diffraction is contained in the following query:—"Do not the rays which differ in refrangibility differ also in flexibility and are they not by their different inflexions separated from one another, so as, after separation, to make the colours in these fringes" [of the diffraction pattern he had previously described]? And again,

"Are not the rays of light in passing by the edge and sides of bodies, bent several times backwards and forwards, with a motion like that of an eel? And do not the three fringes of coloured light arise from three such bendings?" (Q. 3.)

* Cf. the theory of the X-rays.

† Cf. Kepler's laws of Planetary Motion, in particular the third law.

33. But, although Newton worked out these explanations on the basis of the corpuscular theory, he himself had an open mind—"argued the corporeity of light but without any absolute positiveness." He was, moreover, fully cognizant, as we have just stated, of the rival hypothesis and its merits.

Thus, in Q. 13, he proposes an explanation of colour on the wave theory:

"Do not several sorts of rays make vibrations of various bignesses, which according to their bigness excite sensations of various colours, the most refrangible, the shortest vibrations?"

Again, in Q. 17 he puts forward what may be called a combined theory as a possible explanation of the 'Fits' [Art. 31].

But in Q. 28, he points out the crucial difficulties of the wave theory—the rectilineal propagation of light. "If it (light) consisted in pressure or motion, propagated either in an instant or in time, it would bend into the shadow. For pressure or motion cannot be propagated in a fluid in right lines beyond an obstacle, which stops part of the motion but will bend and spread every way into the quiescent medium which lies beyond the obstacle."

Nor was it at all clear how double refraction was to be explained if light consisted in vibrations, like those of sound.

34. It was these difficulties which led him (in Q. 29) to suggest the corpuscular theory and point out how it served to explain the formation of shadows. "Are not the rays of light very small bodies emitted from shining substances? For such bodies will pass through a uniform medium in right lines without bending into the shadow, which is the nature of the rays of light."

35. Previously to this, however, he had proposed a comprehensive hypothesis, which, as a mere hypothesis, he did not think it desirable to include in his treatise, though it was of an older date. In a paper sent to Oldenburg for the Royal Society but withheld from publication at his own request, he thus states his hypothesis:

"Were I to assume an hypothesis, it should be this, if propounded more generally, so as not to assume what light is further than that it is something or other, capable of exciting vibrations of the ether. First, it is to be assumed that there is an ethereal medium, much of the same constitution with air but far rarer, subtiller and more strongly elastic. In the second place, it is to be supposed that the ether is a vibrating medium, like air; only the vibrations, far more swift and minute; those of air made by a man's ordinary voice, succeeding at more than half a foot or a foot distance, but those of ether at less distance than the hundred thousandth part of an inch. And as in air, the vibrations are some larger than others but yet all equally swift, so I suppose the ethereal vibrations differ in bigness but not in swiftness. In the fourth place, therefore, I suppose that light is neither ether nor its vibrating motion but something of a different kind propagated from lucid bodies. They that will may suppose it an aggregate of many peripatetic qualities. Others may suppose it to consist of multitudes of unimaginable small and swift corpuscles of various sizes, springing from shining bodies at great distances, one after the other, but yet without any sensible interval of time....I would suppose it diverse from the vibrations of the ether. Fifthly, it is to be supposed that light and ether mutually act upon one another. It is from this action that reflection and refraction come about. Ethereal vibrations are, therefore, the best means by which such a subtle agent as light can shake the gross particles of solid bodies to heat them."

"And now to explain colour. I suppose that as bodies excite sounds of various tones and consequently vibrations in the air of various bignesses, so when the rays of light, by impinging on the stiff refracting superficies excite vibrations in the ether, these rays excite vibrations of various bignesses—the biggest with the strongest colours, reds and yellows; the least with the weakest, blues and violets; the middle with green and a confusion of all with white."

36. This is, practically, the undulatory theory, which in the hands of Huyghens, Young and Fresnel furnished a fairly satisfactory explanation of all the known phenomena of optics. The explanation, however, is only kinematical. As contrasted with this, the corpuscular theory, so far as it goes, has a dynamical basis. As such, it evidently has, *prima facie*, an advantage over the undulatory theory, and, accordingly, deserves further consideration.

37. The first and the simplest objection to the corpuscular theory, which appeared almost conclusive at one time, was based on the fact that light corpuscles appeared to be absolutely without momentum. It was, however, easy to meet this objection by attributing to light corpuscles a minuteness, sufficient to evade any means that was then possessed of detecting their existence by their effects on other bodies. We now know (1) that light does exert an impulsive effect and (2) that corpuscles moving with velocities comparable with that of light are at least as real as the chemical atom.

38. On the other hand, the phenomenon of aberration is much more simply explained on this theory than on that of undulation. When, however, we come to consider the theory of refraction, apart from the very artificial nature of the theory of fits on which its explanation is based, it is found that it is necessary to suppose that the velocity increases with the density of the medium. But experiment shows that the reverse is the case and we have, accordingly, to reject the theory but only in the special form in which it was applied by Newton to the explanation of refraction. For the disagreement might have been due to the fact that the mode and law of interaction between 'gross' matter and corpuscles as postulated by Newton was wrong, but it would not necessarily prove that light is not propagated by means of corpuscles.

39. As to the velocity of corpuscles, modern theory postulates very high velocities for the cathode rays and for particles projected from radium. There was, accordingly, nothing *a priori* absurd in the corpuscular view.

40. But light has two characteristics, not only a velocity of propagation but also a periodicity. Newton must have felt the inadequacy of the corpuscular theory in explaining this—the latter aspect of optical phenomena—and it was on account of this, obviously, that he had to introduce the element of time—the time of a fit—in his explanations and as even that did not seem to account for all known phenomena, he definitely adopted the '*bignesses* of vibrations,' as alone explaining colour. He could not, however, reconcile himself to the undulatory theory in its entirety, because of its failure to explain rectilineal propagation, although in one respect he was more precise than the advocates of the undulatory theory themselves; for he laid down that light was neither ether nor its vibrating motion, but energy, and it was the mode of propagation of this energy through an ethereal medium—for such a medium must be postulated in any case—about which it was necessary to frame a hypothesis.

Section III. Undulatory Theory

41. If it is difficult to see how the element of periodicity (so essentially associated with optical phenomena) enters in the motion of corpuscles (on the corpuscular theory), periodicity enters as a fundamental element in wave motion. There was, thus, an initial justification for the theory which regards the energy of light as due to wave motion. As soon, therefore, as the crucial difficulties propounded by Newton, viz. the phenomenon of rectilineal propagation and of double refraction, were explained on this theory, it served to furnish a complete kinematical explanation of all optical phenomena.

The phenomenon of rectilineal propagation was ultimately explained on Huyghens' principle. The principle, simply, is that each element of a wave surface may be regarded as a source of the disturbance and, therefore, (1) the wave surface at any subsequent instant is the envelope of the wave surfaces, constructed with these elements as sources; (2) the disturbance at any point is the vector-sum of the disturbances due to these elements*. The first of these principles enabled Huyghens to give a construction and a complete (kinematical) explanation of

* The weight to be attached to each element was naturally left doubtful. This was supplied by Stokes [Art. 178].

refraction and reflection. It further served (on the assumption that the wave surface inside a uniaxial crystal, like iceland spar, consists of a sphere and a spheroid) to explain the phenomenon of double refraction, also, in such crystals.

42. The second principle would, if properly interpreted have led to the proof of the rectilineal propagation, but it was not till quite a hundred years later that this was successfully accomplished by Young and Fresnel. For the second is in reality the principle of interference, from which it follows that since the wave-length of light is infinitely small, luminous effect at any point must be due to a small portion of a whole wave front, and, therefore, light must, *as a rule*, be propagated in right lines but that it should (as it does, as a matter of fact) bend round apertures,—even as Newton argued it ought to,—if the aperture is sufficiently narrow. This mode of regarding the phenomena also effectively meets the initial difficulty to which the undulatory theory is subject, viz. that it leads to the conclusion that there ought to be shadows of sound as well as those of light. For, inasmuch as the wave-length of light is infinitesimal, as compared with that of sound, sound shadows can be produced only by large obstacles, as we know they actually are.

43. A fairly complete mathematical explanation of the ordinary phenomena of optics, based on the principles of the wave theory propounded by Huyghens and Young, was given by Fresnel, but his explanations were mainly kinematical. Where he attempted a dynamical explanation, that explanation on examination is found to be either arbitrary, or at any rate to involve postulates which it is difficult to interpret.

44. Before proceeding to the consideration of the attempts that have been made at a strictly dynamical explanation, we shall examine the kinematical explanations that the undulatory theory has supplied and to which reference has just been made [Art. 41].

45. If v is the velocity of a periodic disturbance, we have

$$v = a \sin 2\pi \left(\frac{t}{\tau} - \frac{x}{\lambda}\right),$$

where a, the maximum velocity, is proportional to the amplitude, τ, the periodic time, and λ, the wave-length, while x is the distance at which the velocity is v, the velocity at the origin being at the same time $(t) = a \sin 2\pi \frac{t}{\tau}$. Then, it easily follows that the mean kinetic energy of the disturbance

$$= \frac{1}{\tau}\int_t^{\tau+t} \frac{a^2}{2} \sin^2 2\pi \left(\frac{t}{\tau} - \frac{x}{\lambda}\right) dt = \tfrac{1}{4}a^2,$$

the mass of a vibrating element being taken to be equal to unity.

Consider, now, the effect at any point of two disturbances from two origins (sources) at distances x_1, x_2.

If the velocities are in the same direction [Art. 48] (whatever be this direction) the resultant velocity will be

$$v_1 + v_2 = a_1 \sin 2\pi \left(\frac{t}{\tau} - \frac{x_1}{\lambda}\right) + a_2 \sin 2\pi \left(\frac{t}{\tau} - \frac{x_2}{\lambda}\right)$$

$$= A \sin 2\pi \left(\frac{t}{\tau} - \frac{x}{\lambda}\right), \text{ say,}$$

where $$A^2 = a_1^2 + a_2^2 + 2a_1a_2 \cos 2\pi \left(\frac{x_1 - x_2}{\lambda}\right).$$

Therefore, (1) $A = 0$, or the resultant disturbance is zero, i.e. the resultant effect is one of absence of light, if

$$a_1 = a_2,$$

and $$x_1 - x_2 = (2n+1)\frac{\lambda}{2}.$$

(2) Under the same conditions, A has its maximum value or the resultant disturbance gives the maximum of illumination, if $x_1 - x_2 = n\lambda$, where n is an integer.

(3) If $\frac{x_1 - x_2}{\lambda}$ varies irregularly, during the time $n\tau$, i.e. if the disturbance at the origin changes in phase from moment to moment, independently of each other, the mean of

$$\cos 2\pi \frac{x_1 - x_2}{\lambda}$$

during $n\tau$, where n is large,

$$= \frac{1}{n\tau}\int_t^{n\tau+t} \cos 2\pi \frac{x_1 - x_2}{\lambda}\, dt = 0,$$

since $x_1 - x_2$ goes through all possible values during the interval.

In this case $A^2 = a_1^2 + a_2^2$, always.

The third result explains why two lights cannot *interfere*, if they proceed from two different sources. This, indeed, is an experimental fact from which we conclude that vibrations from two sources change their character, independently of each other, from moment to moment.

(1), (2) specify the conditions under which interference can take place.

Exactly the same principle applies when we proceed to explain the effects observed, when light passes through a narrow aperture or a large number of narrow apertures arranged in any manner (a grating) [Art. 180]. This is the phenomena of diffraction.

Generally, therefore, when light from the same source reaches a point after traversing two different paths, we have the phenomena of interference.

46. It has sometimes been argued that the interference effects as shown by the experiments of Young and Fresnel are fatal to the emission theory. It should be noted, however, that in all these cases, there is a redistribution of energy but no loss and, therefore, if we conceive light to be propagated by perfectly elastic particles (which are of the nature of revolving doublets), these by their collision *may* well change the distribution of energy in the way in which it is actually found to do*.

47 The principle of Huyghens, then, and that of interference which is really involved in it, serve to give a complete kinematical explanation of rectilineal propagation, reflection, refraction, interference and diffraction of light.

48. In explaining these, it is unnecessary to define the direction of vibration in relation to the ray. When we come to double refraction, the question of this relation comes naturally to the fore. Let us recall the fact that when a ray of ordinary light is passed through a piece of iceland spar, the emergent light is divided into two rays; one of them is refracted according to the ordinary law of refraction and is called the ordinary ray while the other does not obey this law (the corresponding index of

* Preston, *Theory of Light.*

refraction depending on the direction of the incident ray) and is therefore called the extraordinary ray. It is just conceivable (on some suitable hypothesis), that even if the vibrations of light were longitudinal, the incident ray would be thus decomposed. But when these decomposed rays are again passed through another piece of iceland spar, it is found that, as the principal section of the second crystal is rotated round that of the first [Art. 57], the diverging rays (two for each ray) vary in brightness. The vibrations cannot therefore be entirely longitudinal. It is, moreover, found that when ordinary light from a given source is decomposed by means of a uniaxial crystal like iceland spar, any two of the resulting ordinary rays (or of the extraordinary rays) if allowed to pursue slightly different paths will interfere, but that an ordinary ray will not interfere with an extraordinary ray and *vice versa*. From this, it follows (1) that the vibrations of light cannot be even partially longitudinal, i.e. they must be perpendicular to the ray, (2) the ordinary and the extraordinary rays must each continuously vibrate in or perpendicular to a certain invariable plane and (3) that these planes must be perpendicular to each other.

49. These rays are said to be plane polarised, so that the difference between plane polarised light and ordinary light consists in this that in the latter the vibrations continuously change their azimuths round the ray, while in the former these azimuths remain unchanged. And whenever and through whatever cause, the vibrations have this character impressed on them (whether the light suffers double refraction at the same time or not) they are said to be plane polarised. Thus, double refraction consists not in creating these transverse motions but in decomposing them into two lights, polarised in two planes at right angles to each other, and separating the components.

50. The above mode of analysis of the vibrations constituting light being admitted, the explanation of elliptic and circular*

* The difference between these and ordinary light again consists in this that in the case of the latter, when decomposed in two fixed directions, the elements a, α, b, β continuously change, while in the case of the former, these are invariable.

polarisations as the result of compounding two plane polarised vibrations of suitable types, namely

$$a \sin\left(\frac{2\pi}{\tau} - \alpha\right) \text{ and } b \sin\left(\frac{2\pi}{\tau} - \beta\right),$$

rectangular or oblique, easily follows.

51. In attempting, in the manner of Fresnel, a more detailed explanation of double refraction, we note in the first place that in the case of a uniaxial crystal, the assumption seems legitimate that the wave surface in such a crystal consists of a sphere and spheroid. For it is, *a priori*, evident that luminous vibrations would be propagated in a homogeneous medium in spherical waves, and hence it is natural to admit that in a *doubly-refracting* medium, having an *axial symmetry*, they will be propagated in a sphere and a spheroid. This leads, on Huyghens' principle, to the following construction:

52. Describe a sphere and a spheroid with a point on the refracting surface as centre and the axis of revolution of the spheroid coincident with the axis of the crystal. Through any point in the plane of incidence, belonging to the refracting surface, draw tangent planes to these surfaces. The radii-vectores to the points of contact are the refracted rays and the perpendiculars to the tangent planes represent, on a suitable scale, the velocities of propagation.

53. Let the equations of these surfaces be

$$x^2 + y^2 + z^2 = a^2 \quad \dots\dots\dots\dots\dots\dots(1),$$

and

$$\frac{x^2}{a^2} + \frac{y^2 + z^2}{b^2} = 1 \quad \dots\dots\dots\dots\dots\dots(2).$$

Let also l, m, n define the wave front [Art. 17] and let v_1 be the velocity of propagation of the ordinary ray, l', m', n', v_2 being similar quantities for the extraordinary ray; then, from the definition of the wave surface, we have

$$(l^2 + m^2 + n^2)\, a^2 = v_1^2,$$

and

$$a^2 l'^2 + b^2 (m'^2 + n'^2) = v_2^2,$$

or, $\frac{1}{v_1^2}$, $\frac{1}{v_2^2}$ are the roots of the quadratic in r^2, viz.

$$\left(a^2 - \frac{1}{r^2}\right)\left[l^2\left(a^2 - \frac{1}{r^2}\right) + (m^2 + n^2)\left(b^2 - \frac{1}{r^2}\right)\right] = 0$$

or, $$\frac{l^2}{b^2-\frac{1}{r^2}}+\frac{m^2}{a^2-\frac{1}{r^2}}+\frac{n^2}{a^2-\frac{1}{r^2}}=0, \quad\text{..............(3)}$$

i.e. $\frac{1}{v_1}, \frac{1}{v_2}$ are the axes of the section of the spheroid

$$b^2x^2+a^2(y^2+z^2)=1 \quad\text{....................(4)}$$

by the plane $lx+my+nz=0$.

We conclude, therefore, that the reciprocals of the velocities are equal to the lengths of the axes of the central section of a certain spheroid (4) by $lx+my+nz=0$.

54. Generalizing the above, in a biaxial crystal, it is reasonable to suppose that the surface corresponding to (4) of Art. 53 will be an ellipsoid. Let the equation of this surface be $a^2x^2+b^2y^2+c^2z^2=1$.

Hence, it easily follows, that l, m, n, the direction-cosines of the wave normal, will be connected by the relation

$$\frac{l^2}{v^2-a^2}+\frac{m^2}{v^2-b^2}+\frac{n^2}{v^2-c^2}=0,$$

corresponding to (3) of Art. 53, and the wave surface will be the envelope [Art. 52] of $lx+my+nz=v$, where

$$\frac{l^2}{v^2-a^2}+\frac{m^2}{v^2-b^2}+\frac{n^2}{v^2-c^2}=0$$

and $$l^2+m^2+n^2=1.$$

Further, if λ, μ, ν be the direction-cosines of an axis of the central section of $a^2x^2+b^2y^2+c^2z^2=1$ (principal axis, r), we have

$$v^2=r'^2=a^2\lambda^2+b^2\mu^2+c^2\nu^2 \quad\text{..............................(1)},$$

$$0=l\lambda+m\mu+n\nu \quad\text{..................................(2)},$$

$$0=\frac{l}{\lambda}(b^2-c^2)+\frac{m}{\mu}(c^2-a^2)+\frac{n}{\nu}(a^2-b^2) \quad\text{........(3)}.$$

So far, the work is entirely geometrical and is derived by a reasonable generalization of Huyghens' assumption that in a uniaxial crystal, the wave surface consists of a sphere and a spheroid. It remains now to see whether a dynamical interpretation of the equations (1), (2), (3) is possible.

Now, the velocity of propagation of a transverse wave in a stretched string or cord is $\sqrt{\frac{\tau}{\rho}}$, where τ is the tension of the string or stress per unit length, and ρ, the density.

If, therefore, we can postulate the operation of elastic forces of this type in an elastic medium, a dynamical interpretation of (1), (2) and (3) is possible.

In order to see how the requisite stress may be accounted for, let two of the axes of reference be the axes of the crystal and let OP = the displacement of a particle at any time

$$= \xi = (\xi\lambda,\ \xi\mu,\ \xi\nu).$$

If this is supposed to bring into play a force of restitution

$$= (a^2\rho\xi\lambda,\ b^2\rho\xi\mu,\ c^2\rho\xi\nu)\ \text{(say)} \quad \ldots\ldots\ldots\ldots(4),$$

the resolved part of this force along λ, μ, ν, per unit displacement

$$= \rho\,(a^2\lambda^2 + b^2\mu^2 + c^2\nu^2) \ldots\ldots\ldots\ldots\ldots\ldots(5).$$

(1) can, therefore, be interpreted, if we assume that the force of restitution, called into play, is of the above type (4) of which the only part that is effective in producing vibration inside the crystal is that contemplated in (5). It follows further that a, b, c are the principal wave-velocities.

(2), obviously, means that the direction of velocity is perpendicular to the direction of the wave-normal.

(3) is equivalent to $\begin{vmatrix} a^2\lambda & b^2\mu & c^2\nu \\ \lambda & \mu & \nu \\ l & m & n \end{vmatrix} = 0,$

which means that the wave-normal, the displacement and the force of restitution are in the same plane. In other words, only those displacements are effective which give rise to stress in the plane containing the wave-normal and the displacement, i.e. those that give rise to stress whose component along the wave-front is along the direction of displacement.

55. It is hardly necessary to enter on a detailed examination of hypotheses which were actually* made, in order to obtain a particular analytical result. As, however, the final result deduced from them has been verified by experiment, it would be as well to examine them briefly.

The hypotheses are:

(a) The force of restitution is proportional to displacement (the constant of proportionality being the product of density and a factor depending on direction).

If, however, the ethereal medium inside a crystal behaves like an elastic solid, the force should be proportional to strain

* Whittaker, *A History of Theories of Ether and Electricity*, p. 125.

[Art. 70]. On the other hand, if the equation of motion is that of an electron [Ch. IV], this objection does not apply.

(*b*) Only the resolute of the force along the direction of displacement is effective. This is obviously dynamically unsound, unless we hypothecate this as a property of the medium. (Cf. the incompressible or labile ether of the elastic solid theory.)

(*c*) The vibrations are perpendicular to the direction of propagation.

(*d*) The wave-normal, the displacement and the force of restitution are in the same plane. This does not seem to have any justification, unless we regard it as defining the property of the medium, due to structure, etc.

The phenomenon of conical refraction was at one time regarded as a singular verification of Fresnel's theory, but it only indicates that the wave surface should be a surface of two sheets, having conical points, and cannot tell us anything as to the special hypotheses made*.

56. Fresnel's wave surface†.

Let OP be perpendicular to the tangent plane to an ellipsoid S (centre O), say, at the point Q, and OP' the consecutive perpendicular. Let

$$S \equiv \frac{x^2}{a^2} + \frac{y^2}{b^2} + \frac{z^2}{c^2} - 1 = 0$$

[Art. 54], and let S' be reciprocal to S.

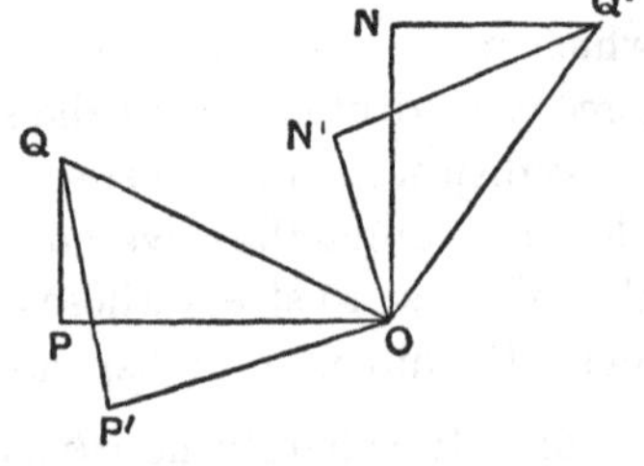

Also let ON be perpendicular to the central section of S' containing OP and let $ON = OP$, and finally let P' N' be consecutive points to P, N, the locus of Q' being the envelope of the plane perpendicular to ON.

Then OQ' is equal and perpendicular to OQ.

Now, if p = perpendicular to the tangent plane (λ, μ, ν) to S,

$$p^2 = a^2\lambda^2 + b^2\mu^2 + c^2\nu^2 = v^2,$$

if p is drawn in the direction of displacement. Then $ON = v$, and as it is drawn perpendicular to the central section of S', the wave surface is the envelope of planes perpendicular to ON [Art. 54].

* *B. A. Report*, 1862. † MacCullagh, *Collected Works*, I.

Hence, this surface is the locus of points at the extremities of lines drawn perpendicular and proportional to the radius vector (OQ) of the ellipsoid S, corresponding to the perpendicular OP, in the direction of displacement, to the tangent plane to S.

Also $$S' \equiv a^2x^2 + \ldots + \ldots - 1 = 0.$$

Now since OP is one of the axes of the central section of S', OQ will be one of the axes of the central section of S and OQ' will be the normal to that section (say, l, m, n).

Therefore, writing $OQ = OQ' = r$, we have

$$\frac{l^2}{\frac{1}{a^2} - \frac{1}{r^2}} + \ldots + \ldots = 0,$$

and $lr = x$, etc., if x, y, z are the coordinates of Q'. Whence the equation of the wave surface is

$$\frac{a^2x^2}{r^2 - a^2} + \ldots + \ldots = 0,$$

In the case of a uniaxial crystal, it is easy to verify that the surface becomes a sphere and a spheroid. For, since the wave surface in the general case is

$$\frac{a^2x^2}{r^2 - a^2} + \ldots + \ldots = 0,$$

if $b = c$, we have

$$(r^2 - b^2)\,[a^2x^2 + b^2(y^2 + z^2) - a^2b^2] = 0.$$

57. The hypotheses [Art. 55], when applied to the particular case of a uniaxial crystal, will amount to the following. Let OX be the axis of the crystal, OZ, perpendicular to OX in the wave front, and OP, the direction of displacement; let also $\angle ZOP = \alpha$ and the displacement $(OP) = \sigma$.

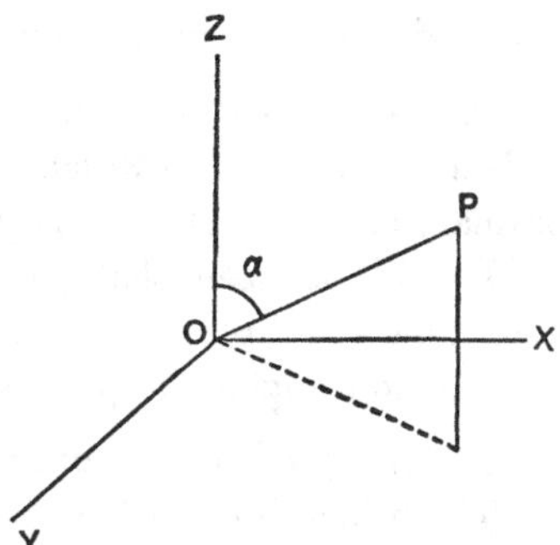

Then, the resolved part of the displacement along OZ is $\sigma \cos \alpha$.

Since the force of restitution is in the direction of displacement from symmetry, it will give rise to a ray along the axis. And every incident ray gives rise to a refracted ray along the axis, no matter what its direction is This constituent is therefore the 'ordinary ray.'

That is, the vibrations in the ordinary ray are perpendicular to the axis of the crystal.

Consider, now, the following experiment*.

A ray of light (unpolarised) is passed through a spar in the principal section which is vertical. If, now, we cut off the extraordinary ray and let the ordinary ray of intensity O pass through a second spar in the principal plane, we get two images of intensities O', E', given by $O' = O \cos^2 \alpha$, $E' = O \sin^2 \alpha$†, where α is the inclination of the second principal section to the vertical. Therefore if $\alpha = 0$, $O' = O$, i.e the ordinary ray is in *fixed* relation to the principal plane; it *is said* to be polarised in that plane. Now, since the vibrations of the ordinary ray, on Fresnel's theory, are perpendicular to the axis, Fresnel's theory involves the supposition that the vibrations of plane polarised light are perpendicular to the plane of polarisation.

It is easy to see that this mode of regarding the phenomena is consistent with experimental results. For this, let us discuss the experiment cited above.

A vibration given by $A \cos \frac{2\pi t}{\tau}$ is incident on the second crystal, inclined at α to OZ (say) drawn perpendicular to the ray and the axis. It can be resolved into

$$A \cos \alpha \cos \frac{2\pi t}{\tau} \text{ along } OZ \text{ (ordinary ray)}$$

and $$A \sin \alpha \cos \frac{2\pi t}{\tau} \text{ perpendicular to } OZ \text{ (extraordinary ray).}$$

Then, the intensities are as $\cos^2 \alpha : \sin^2 \alpha$. But this is the law of Malus, according to which the ratios of the intensities of the ordinary and the extraordinary rays are as $\cos^2 \alpha : \sin^2 \alpha$.

It will be seen that this experiment does not uniquely determine the direction of vibration of plane polarised light, in relation to the plane of polarisation.

58. Fresnel's theory of reflection and refraction.

The principles on which the theory is based are:

(1) The vibrations are perpendicular to the plane of polarisation.

* Jamin et Bouty, *Cours de Physique*, Tome III, 3. † Law of Malus.

(2) The density is different in different media, but rigidity the same and, therefore, the velocity $\left(=\sqrt{\frac{\text{rigidity}}{\text{density}}}\right)$ varies inversely as the square root of the density.

(3) The tangential displacement is continuous.

(4) The phase is unchanged at reflection and refraction.

(5) The total energy in the reflected and refracted wave is equal to the energy of incident wave.

Assuming (1), Fresnel's theory leads to the conclusion that the intensity of natural light reflected is*

$$\frac{1}{2}\left[\frac{\sin^2(i-r)}{\sin^2(i+r)}+\frac{\tan^2(i-r)}{\tan^2(i+r)}\right],$$

the first term giving the intensity of light polarised in the plane of incidence, while the second, that perpendicular to this plane. When, therefore, $i+r=90°$, i.e. at the polarising angle, we get only the first term and light is polarised entirely in the plane of incidence.

This may serve as the basis of an experimental method of determining the plane of polarisation and also provides (on Fresnel's theory) the basis of the following definition, viz.:—the plane of polarisation is the plane in which light is polarised by reflection.

59. Now, we have seen that when light is polarised in its passage through iceland spar, the plane of polarisation of the ordinary ray is the principal plane (of the spar) containing the ray. The ordinary ray also follows the ordinary law of refraction, viz. the plane of incidence is the same as the plane of refraction. Therefore, the plane of incidence is the same as the plane of polarisation. The principle (1) is thus seen to be consistent with the hypothesis on which double refraction is explained.

(2), (3). These are also the hypotheses of the 'Labile' ether theory [Art. 72].

(4). This is inconsistent with experiment. There is always change of phase at reflection and refraction.

* Jamin et Bouty, Tome III, 3.

60. Again, Fresnel's theory leads to the conclusion that for light polarised perpendicularly to the plane of incidence, the intensity of reflected light is $\frac{\tan^2(i-r)}{\tan^2(i+r)}$*.

Therefore when $i+r=90°$, i.e. at the polarising angle, there should be no reflected light. Experiment shows however that this is not the case. Lorentz has shown that a transition layer of ether of variable density will account for this as well as the change of phase [Art. 58].

Now, from the continuity of energy [(5), Art. 58], we get

$$S\lambda\rho a_1^2 = S\lambda\rho a_2^2 + S'\lambda'\rho' a'^2,$$

where S, S' = the areas of corresponding elements of the wave fronts in the two media,

λ, λ' = wave-lengths,

ρ, ρ' = densities of the media, and

(a_1, a_2, a') = amplitudes of incident, reflected and refracted vibrations,

$$\therefore\ a_1^2 - a_2^2 = a'^2\frac{\rho'}{\rho}\cdot\frac{\cos r}{\cos i}\frac{1}{n} \quad\text{...........(1)},$$

since $$\frac{\sin i}{\sin r} = \frac{V}{V'} = n = \frac{\lambda}{\lambda'} \text{ and } \frac{S}{S'} = \frac{\cos i}{\cos r},$$

where n = index of refraction, V, V' = velocities of propagation in the two media. Also

$$\frac{\rho'}{n\rho} = n = \frac{\sin i}{\sin r} \quad\text{......................(2)},$$

if the rigidity is the same in both media.

But if both the tangential and normal displacements are continuous, we must have

$$a_1 - a_2 = a'\frac{\cos r}{\cos i} \quad\text{.....................(3)},$$

$$a_1 + a_2 = a'\frac{\sin r}{\sin i} \quad\text{.....................(4)}.$$

$\therefore$ from (2), (3) and (4) $a_1^2 - a_2^2 = a'^2\cdot\frac{\cos r}{n\cos i}$ ....(5),

$\therefore\ \rho = \rho'$, from (1) and (5), which is absurd.

This shows that, on Fresnel's theory, both the displacements cannot be regarded as continuous.

* Jamin et Bouty, Tome III, 3.

CHAPTER II

ELASTIC SOLID THEORY

SECTION I. THEORY OF ELASTICITY. REFLECTION REFRACTION

61. We have seen that the method of Fresnel is not strictly dynamical*. Now in order to obtain a strictly dynamical explanation of optical phenomena, it is necessary to postulate certain properties of the medium which takes part in the propagation of the vibrations constituting light.

The simplest hypothesis that we can make in order to explain the propagation of the vibrations of light would be to suppose that the medium is an elastic medium, and since these vibrations are transverse, it must possess the property of resisting shearing strain. Now, the only bodies that we know, possessing such properties, are elastic *solids*. The postulation of the transfer of the transverse vibrations of light by a medium, in virtue of its elastic properties, therefore, amounts to the supposition that such a medium behaves as an elastic solid. It is the model, we naturally conceive, in order to explain the peculiar features of the phenomena. The difficulty of conceiving such a model, however, is obvious—the difficulty, namely, of associating the interstellar space with the properties of an elastic solid. But it is well-known that a body behaves differently under different circumstances of motion. In the case of steady motion of bodies through air, for instance, it behaves very nearly like a perfect fluid—but perfectly compressible; yet, when the rapid vibratory motion of sound is imposed on it, it offers resistance

* Or, rather, it is dynamical, on the basis of special hypotheses.

to compression, like an elastic body Similarly, a medium may have all the properties of a perfect fluid for steady motion like that of the heavenly bodies, but may offer resistance to shear, when extremely rapid transverse vibrations, constituting light, are impressed on it.

62. The following remarks of Stokes [*Papers*, Vol. II, p. 12] clearly set forth the point of view on which the elastic solid theory is based:

"Suppose a small quantity ot glue dissolved in a little water so as to form a stiff jelly This forms in fact an elastic solid. It may be constrained, and will resist constraint and return to its original form when the constraint is removed, by virtue of its elasticity; if constrained too much, it will break. Suppose the quantity of water in which the glue is dissolved is gradually increased till it is only glue-water. At last, it will be so far fluid as to mend itself again, as soon as it is dislocated. Yet, there seems hardly sufficient reason for supposing that at a certain stage of the dilution, the tangential force, whereby it resists constraint, ceases all of a sudden. In order that the medium should not be dislocated and therefore should have to be treated as an elastic solid, it is only necessary that the amount of constraint should be very small. The medium would however be what we should call a fluid, as regards the motion of solid bodies through it The velocity of propagation of normal vibrations in our medium will be nearly the same as that of sound in water; the velocity of propagation of transverse vibrations depending as it does on the tangential elasticity will become very small. Conceive, now, a medium having similar properties, but incomparably rarer than air, and we have a medium such as we conceive ether to be—a fluid as regards the motion of the earth through it, an elastic solid as regards the small vibrations which constitute light. The sluggish transverse vibrations of our thin jelly are in the case of the ether replaced by vibrations propagated with a velocity of nearly 200,000 miles per second. We should expect *a priori* the velocity of propagation of normal vibrations to be incomparably greater."

63*. If u, v, w be the displacements of a point (x, y, z) of an elastic solid, then $x+u$, $y+v$, $z+w$ will be its displaced coordinates.

The displaced coordinates of $x+\Delta x$, etc. will be

$$x+\Delta x+u+\frac{\partial u}{\partial x}\Delta x, \text{ etc.};$$

$\therefore$ the new lengths of Δx, Δy, Δz will be

$$\Delta x\left(1+\frac{\partial u}{\partial x}\right), \text{ etc.}$$

$\therefore$ the linear extension in the direction of x is

$$\text{Lt}\,\frac{\Delta x\left(1+\frac{\partial u}{\partial x}\right)-\Delta x}{\Delta x}=\frac{\partial u}{\partial x}\equiv e_{xx}, \text{ say,}$$

and the voluminal expansion

$$=\left(\frac{\partial u}{\partial x}+\frac{\partial v}{\partial y}+\frac{\partial w}{\partial z}\right)\equiv\Delta, \text{ say.}$$

64. Let a rectangle $ABDC$ be held fixed by the edge AB and pulled by a force along CD, AC and BD being displaced to AC', BD'.

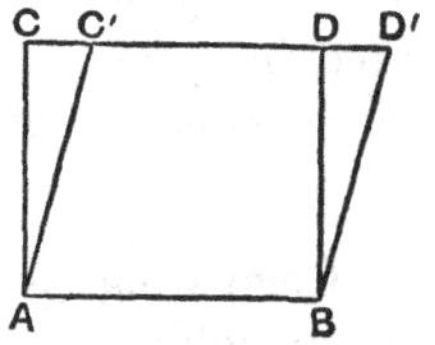

C C′ D D′ A B

Then

$$\theta=\frac{CC'}{AC}=\text{Lt}\,\frac{\Delta u}{\Delta y}=\frac{\partial u}{\partial y},$$

if AB is taken along x, and AC along y and equal to Δy, while $\theta=$ the circular measure of the $\angle CAC'$.

Hence, for a rectangular parallelepiped, subjected to tangential stress, with the plane $z=0$ fixed, the strain

$$=\frac{\partial v}{\partial x}+\frac{\partial u}{\partial y}\equiv e_{xy}, \text{ say};$$

e_{xy} is called a shearing strain.

65. If U, V, W be the components of displacement of any point $(x+\xi, y+\eta, z+\zeta)$, then by Taylor's Theorem,

$$U=u+\frac{\partial u}{\partial x}\xi+\frac{\partial u}{\partial y}\eta+\frac{\partial u}{\partial z}\zeta, \text{ neglecting square terms};$$

* Love's *Elasticity*.

∴ the *relative* displacement in the direction of x

$$= \frac{\partial u}{\partial x}\xi + \frac{\partial u}{\partial y}\eta + \frac{\partial u}{\partial z}\zeta$$

$$= \frac{\partial u}{\partial x}\xi + \frac{1}{2}\left(\frac{\partial u}{\partial y} + \frac{\partial v}{\partial x}\right)\eta + \frac{1}{2}\left(\frac{\partial u}{\partial y} - \frac{\partial v}{\partial x}\right)\eta$$

$$+ \frac{1}{2}\left(\frac{\partial u}{\partial z} + \frac{\partial w}{\partial x}\right)\zeta + \frac{1}{2}\left(\frac{\partial u}{\partial z} - \frac{\partial w}{\partial x}\right)\zeta$$

$$= e_{xx}\xi + \tfrac{1}{2}e_{xy}\eta + \tfrac{1}{2}e_{xz}\zeta - \omega_z\eta + \omega_y\zeta,$$

provided we take $\frac{\partial v}{\partial x} - \frac{\partial u}{\partial y} \equiv 2\omega_z$, etc.....................(1)

and ξ, η, ζ small.

Therefore, the relative displacement in the direction of x

$$= \frac{\partial \phi}{\partial \xi} - \omega_z\eta + \omega_y\zeta \quad \text{..................(2)},$$

where $2\phi = e_{xx}\xi^2 + e_{yy}\eta^2 + e_{zz}\zeta^2 + e_{yz}\eta\zeta + e_{zx}\zeta\xi + e_{xy}\xi\eta$.

But when a rigid body is in motion and its displacements are referred to axes meeting at a fixed point O but moving in space with angular velocities θ_1, θ_2, θ_3 about themselves, then, if the displacement of any point (x, y, z) in the direction of x fixed in space is X_1, and its displacement in the direction of the moving axis of x is X,

$$X_1 = X - \theta_3 y + \theta_2 z^* \text{..................(3)}.$$

Comparing (2) and (3), we conclude that the displacement at a neighbouring point $(x+\xi, y+\eta, z+\zeta)$ relative to (x, y, z) is a certain linear displacement together with a displacement due to rotations of the axes of reference, the angular displacements of the axes being ω_x, ω_y, ω_z, given by (1).

These are called molecular rotations, for obvious reasons.

66. At any point, let the elastic stresses or forces per unit area be X_x, X_y, etc., where

X_x = stress in direction of x perpendicular to $x = 0$;

X_y = tangential stress in the direction of x, in the plane $y = 0$;

X_z = tangential stress in the direction of x, in the plane $z = 0$, etc.

* Routh's *Rigid Dynamics*, Vol. II. Ch. I.

Considering, now, the equilibrium of an elementary cube, not acted upon by external forces, we have by taking moment about the axis of z through the C.G. of the cube

$$(X_y - Y_x')\frac{a^3}{2} = 0,$$

where Y_x' is the value of Y_x at the opposite edge and a the length of a side; but

$$Y_x' = Y_x + a\frac{\partial Y_x}{\partial x} + \ldots.$$

Therefore when a is infinitesimal, we have $X_y = Y_x$, ultimately.

67. Again resolving along the axis of x, we have

$$-X_x + X_x' - X_y + X_y' - X_z + X_z' = 0,$$

or $$-X_x + \left(X_x + \frac{\partial X_x}{\partial x}a\right) - X_y + \left(X_y + \frac{\partial X_y}{\partial y}a\right)$$
$$- X_z + \left(X_z + \frac{\partial X_z}{\partial z}a\right) = 0,$$

i.e. $$\frac{\partial X_x}{\partial x} + \frac{\partial X_y}{\partial y} + \frac{\partial X_z}{\partial z} = 0,$$

and two similar equations.

The corresponding equations of motion obviously are

$$\rho\ddot{u} = \frac{\partial X_x}{\partial x} + \frac{\partial X_y}{\partial y} + \frac{\partial X_z}{\partial z} \quad \text{etc.}$$

68. Let such a cube be strained (under no body-forces), the strain being unaccompanied by loss or gain of heat.

If δW = the increase of potential energy per unit volume of the elementary volume $dx\,dy\,dz$ under displacements δu, etc., then

$$\delta W\,dx\,dy\,dz$$
$$= \left[-X_x dy\,dz\,\delta u + \left(X_x + \frac{\partial X_x}{\partial x}dx\right)dy\,dz\left(\delta u + \frac{d\delta u}{dx}dx\right)\right] + \ldots + \ldots$$
$$= \left[\left(X_x\frac{d\delta u}{dx} + \frac{\partial X_x}{\partial x}\delta u\right) + \ldots + \ldots\right]dx\,dy\,dz,$$

rejecting terms of higher orders.

$$\delta W = X_x\delta e_{xx} + \ldots + \ldots;$$

since $\dfrac{\partial X_x}{\partial x} + \ldots + \ldots = 0$, etc. and $\dfrac{d\delta u}{dx} = \delta\left(\dfrac{du}{dx}\right) = \delta e_{xx}$.

$$\therefore \quad \frac{\partial W}{\partial e_{xx}} = X_x, \text{ etc.}$$

W is called the 'strain energy function,' and the equations of motion [Art. 68] become

$$\rho\ddot{u} = \frac{\partial}{\partial x}\left(\frac{\partial W}{\partial e_{xx}}\right) + \frac{\partial}{\partial y}\left(\frac{\partial W}{\partial e_{xy}}\right) + \frac{\partial}{\partial z}\left(\frac{\partial W}{\partial e_{xz}}\right), \text{ etc.}$$

69. Again, since X_x, etc. are functions of e_{xx}, etc., we conclude that W is a function of e_{xx}, etc., which can be expressed in the form

$$\psi_0 + \psi_1 + \psi_2 + \dots,$$

where ψ_0 is a constant, ψ_1, a linear function of e_{xx}, etc., and so on. If, now, powers of e_{xx}, etc. beyond the second are neglected and the medium is free from stress when in equilibrium. i.e. $\psi_1 = 0$, we have, taking $\psi_0 = 0$,

$$W = \psi_2.$$

70. Let a cube [side $ABCD = 1^2$] of a homogeneous isotropic solid be subjected to a pull ($=1$) at one of the faces perpendicular to BC ($=1$). Let the plane $ABCD$ become strained into $A'B'C'D'$, through the action of this force.

Let ϵ be the angle between the initial and final positions of the diagonal plane, i.e. $\angle BGB'$.

Then the figure being a section parallel to $ABCD$ (or $ABCD$ itself)

$$\tan\left(\frac{\pi}{4} - \epsilon\right) = \tan B'OF = \frac{A'B'}{A'D'}.$$

But $\qquad AB - A'B' = \text{lateral contraction} \quad \dots\dots\dots\dots(1),$

and $\qquad A'D' - AD = \text{longitudinal extension} \quad \dots\dots\dots(2),$

while $\qquad \sigma = \text{Poisson's ratio} = \dfrac{\text{lateral contraction}}{\text{longitudinal extension}} \quad \dots(3),$

and $\qquad \dfrac{\text{longitudinal stress}}{\text{longitudinal strain}} = E = \dfrac{1}{\dfrac{A'D' - AD}{AD}} \quad \dots\dots(4).$

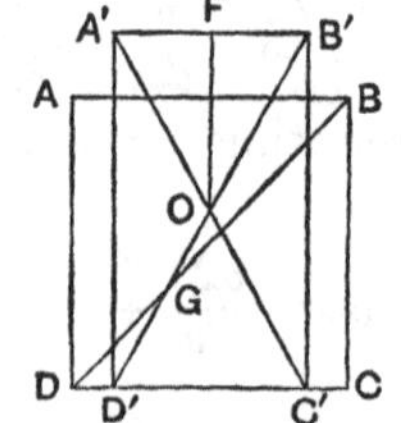

Now, since $AD = 1$, $A'D' = 1 + \dfrac{1}{E}$, from (4),

while $AB - A'B' = \dfrac{\sigma}{E}$, from (1), (3) and (4);

$$\therefore \quad \frac{1-\epsilon}{1+\epsilon} = \frac{1 - \dfrac{\sigma}{E}}{1 + \dfrac{1}{E}},$$

where E is a constant by Hooke's law and is called Young's modulus.

$$\therefore \quad 2\epsilon = \frac{1+\sigma}{E}, \text{ (ϵ being very small)},$$

and the shear on the diagonal $= 2\epsilon = \dfrac{1+\sigma}{E}$ corresponding to a tangential pull per unit area over the diagonal plane equal to $\frac{1}{2}$ (since the resolved part of the pull along the diagonal $= \dfrac{1}{\sqrt{2}}$, and its area is $\sqrt{2}$). Also the coefficient of rigidity is by definition $= \dfrac{\text{shearing stress}}{\text{shearing strain}}$. If we call this μ, we have

$$\mu = \frac{E}{2(1+\sigma)} \quad \text{and} \quad X_y = \mu e_{xy}.$$

Let stresses X_x Y_y, Z_z be applied to the faces of an elementary parallelepiped. Then, since the total extension = longitudinal extension − *lateral* contraction in the same direction, we have, by Hooke's law,

$$e_{xx} = \frac{1}{E}[X_x - \sigma(Y_y + Z_z)], \text{ etc.};$$

$$\therefore \quad E\Delta = (X_x + Y_y + Z_z)(1-2\sigma), \quad \text{[Art. 63]}$$

$$\text{i.e. } X_x = \left(\frac{\sigma}{1-2\sigma}\Delta + e_{xx}\right)\frac{E}{1+\sigma}.$$

Putting $$\frac{E\sigma}{(1+\sigma)(1-2\sigma)} = \lambda,$$

we have $$X_x = \lambda\Delta + 2\mu e_{xx}$$

$$\therefore \quad 2W = (\lambda + 2\mu)\Delta^2 + \mu(e_{xy}^2 + e_{xz}^2 + e_{zy}^2 - 4e_{xx}e_{yy} - \ldots - \ldots).$$

This can be written

$$(\lambda + \tfrac{2}{3}\mu)\Delta^2 + \tfrac{2}{3}\mu\{(e_{xx} - e_{yy})^2 + \ldots + \ldots\} + \mu e_{xy}^2 + \ldots + \ldots :$$

This is always positive if $\lambda + \frac{2}{3}\mu$ be positive and μ, positive. Hence, Δ may be $= 0$, consistently with stability [Art. 72].

71. The expression for W is capable of an interesting transformation: Thus, we have, if S is the bounding surface of the region to which W refers,

$$\int \frac{\partial v}{\partial y}\frac{\partial w}{\partial z}\,dx\,dy\,dz = \left[\int v\frac{\partial w}{\partial z}\,dx\,dz\right]_{\text{bet. limits}} - \int v\frac{\partial}{\partial y}\left(\frac{\partial w}{\partial z}\right)dx\,dy\,dz$$

$$= \int mv\frac{\partial w}{\partial z}\,dS - \int v\frac{\partial}{\partial z}\left(\frac{\partial w}{\partial y}\right)dx\,dy\,dz.$$

Again

$$\int \frac{\partial v}{\partial z}\frac{\partial w}{\partial y}\,dx\,dy\,dz = \int n v \frac{\partial w}{\partial y}\,dS - \int v \frac{\partial}{\partial z}\left(\frac{\partial w}{\partial y}\right) dx\,dy\,dz,$$

where l, m, n are the direction-cosines of the normal to dS;

$$\therefore\quad \int \frac{\partial v}{\partial y}\,\frac{\partial w}{\partial z}\,dx\,dy\,dz = \int \frac{\partial v}{\partial z}\frac{\partial w}{\partial y}\,dx\,dy\,dz + \int v\left(m\frac{\partial w}{\partial z} - n\frac{\partial w}{\partial y}\right) dS.$$

72. Provided, therefore, the surface integral vanishes, i.e.

either $v = 0$, or $\quad m\dfrac{\partial w}{\partial z} - n\dfrac{\partial w}{\partial y} = 0$, etc. ....(1),

at the bounding surface, we have

$$\int \frac{\partial v}{\partial z}\,\frac{\partial w}{\partial y}\,dx\,dy\,dz = \int \frac{\partial v}{\partial y}\frac{\partial w}{\partial z}\,dx\,dy\,dz.$$

Accordingly,

$$\int 2W\,dx\,dy\,dz = \int[(\lambda + 2\mu)\,\Delta^2 + 4\mu(\omega_x^2 + \omega_y^2 + \omega_z^2)]\,dx\,dy\,dz \ldots(2).$$

This shows* that positive work is needed to bring the solid to the strained condition, typified by u, v, w, from its unstrained equilibrium. Therefore, the position of unstrained equilibrium is stable.

Thus, $\lambda + 2\mu$ *may be* zero *consistently with stability.* If we assume $\lambda + 2\mu = 0$, we get the labile-ether theory of Lord Kelvin, which involves the conditions (1), viz., that at the bounding surface, the displacement is either zero or connected by a special relation. The first may be satisfied, if the surface extends to infinity.

If $\Delta = 0$, or the medium is incompressible, while the above surface conditions are satisfied, we have, from (2),

$$\int 2W\,dx\,dy\,dz = 4\int \mu\,(\omega_x^2 + \omega_y^2 + \omega_z^2)\,dx\,dy\,dz.$$

The medium is, then, said to be rotationally elastic.

73. Consider, now, the general expression

$$2W = (\lambda + 2\mu)\,\Delta^2 + 4\mu\,(\omega_x^2 + \omega_y^2 + \omega_z^2)$$
$$\qquad + 2\mu\left[\left(\frac{\partial w}{\partial y}\cdot\frac{\partial v}{\partial z} - \frac{\partial v}{\partial y}\cdot\frac{\partial w}{\partial z}\right) + \ldots + \ldots\right].$$

If δ is the operator of the calculus of variation,

$$\delta\int\left(\frac{\partial w}{\partial y}\cdot\frac{\partial v}{\partial z} - \frac{\partial v}{\partial y}\cdot\frac{\partial w}{\partial z}\right) dx\,dy\,dz$$

* Lord Kelvin, *Phil. Mag.* Vol. XXVI, p. 417.

$$=\int\left[\left(\frac{\partial w}{\partial y}\cdot\frac{\partial\,\delta v}{\partial z}-\frac{\partial w}{\partial z}\cdot\frac{\partial\,\delta v}{\partial y}\right)+\left(\frac{\partial v}{\partial z}\cdot\frac{\partial\,\delta w}{\partial y}-\frac{\partial v}{\partial y}\cdot\frac{\partial\,\delta w}{\partial z}\right)\right]dxdydz$$

$$=\int\left[\left(n\frac{\partial w}{\partial y}-m\frac{\partial w}{\partial z}\right)\delta v+\left(m\frac{\partial v}{\partial z}-n\frac{\partial v}{\partial y}\right)\delta w\right]dS$$

(l, m, n being direction-cosines of the normal at dS)
= surface integrals only.

74. Applying the formula

$$\delta\int(T-W)\,dtdxdydz=0,$$

and remembering that

$$\int T\,dxdydz=\tfrac{1}{2}\int\rho\,(\dot{u}^2+\dot{v}^2+\dot{w}^2)\,dxdydz$$

(T being the kinetic energy per unit volume of the [solid] medium and ρ the density), we get the equations of motion (under no body forces), viz.

$$\tfrac{1}{2}\rho\,\delta\left[\int(\dot{u}^2+\dot{v}^2+\dot{w}^2)\,dtdxdydz\right]$$
$$-\delta\int\left[\tfrac{1}{2}(\lambda+2\mu)\,\Delta^2+2\mu\,(\omega_x^2+\omega_y^2+\omega_z^2)\,dtdxdydz\right]=0,$$

since the remaining terms of W involve surface integrals only—for a homogeneous and isotropic body.

Now (t_1 and t_2 being two specified moments)

$$\tfrac{1}{2}\delta\int\dot{u}^2\,dtdxdydz=\left[\int\dot{u}\,\delta u\,dxdydz\right]_{t_1}^{t_2}$$
$$-\int dt\,\ddot{u}\,dxdydz\,\delta u\;\ldots\ldots(1),$$

and $$\frac{1}{2}\delta\int\Delta^2\,dt\,dxdydz=\int dt\,\Delta\left(\frac{d}{dx}\,\delta u+\ldots+\ldots\right)dxdydz$$

$$=\int dt\,\Delta\,(l\delta u+m\delta v+n\delta w)\,dS$$
$$-\int dt\left(\frac{\partial\Delta}{\partial x}\,\delta u+\frac{\partial\Delta}{\partial y}\,\delta v+\frac{\partial\Delta}{\partial z}\,\delta w\right)dxdydz\;\ldots\ldots(2),$$

while $$\delta\int\omega_x^2\,dtdxdydz=\int dt\,\omega_x\,\delta\left(\frac{\partial w}{\partial y}-\frac{\partial v}{\partial z}\right)dxdydz$$

$$=\int dt\,\omega_x\left(\frac{\partial}{\partial y}\,\delta w-\frac{\partial}{\partial z}\,\delta v\right)dxdydz$$

$$=\int dt\,\omega_x\,(m\,\delta w-n\,\delta v)\,dS$$
$$-\int dt\left[\frac{\partial\omega_x}{\partial y}\,\delta w-\frac{\partial\omega_x}{\partial z}\,\delta v\right]dxdydz\;\ldots\ldots\ldots\ldots(3).$$

Now the volume integrals refer to each element of volume, and the surface integrals refer to the boundaries. The latter will therefore yield the boundary conditions, while the former will similarly give the equations of motion of each element. Hence, collecting the terms in (1), (2), (3) involving only the volume integrals and remembering* that the terms multiplying δu, δv, δw should be separately zero (these being arbitrary), and that the terms in square brackets in (1) vanish, under the usually imposed condition that the initial and final circumstances of the motion (i.e. at t_1 and t_2) are prescribed, we get three equations, viz.

$$(\lambda + 2\mu)\frac{\partial \Delta}{\partial x} - 2\mu\left(\frac{\partial \omega_z}{\partial y} - \frac{\partial \omega_y}{\partial z}\right) = \rho\ddot{u}, \text{ etc. } \quad \ldots\ldots\ldots(4).$$

Differentiating the first with regard to x, the second with regard to y and the third with regard to z and adding, we get

$$\rho\ddot{\Delta} = (\lambda + 2\mu)\nabla^2\Delta \quad \ldots\ldots\ldots\ldots\ldots\ldots\ldots(5).$$

Similarly, differentiating the first with regard to y and the second with regard to x and subtracting, we get

$$\rho\ddot{\omega}_z = \mu\nabla^2\omega_z \quad \ldots\ldots\ldots\ldots\ldots\ldots\ldots\ldots(6).$$

In the same way, we have

$$\rho\ddot{\omega}_x = \mu\nabla^2\omega_x, \quad \rho\ddot{\omega}_y = \mu\nabla^2\omega_y.$$

These are evidently equations of wave propagation.

75. Let

$$\Delta = f(lx + my + nz - c_1 t) \quad \ldots\ldots\ldots\ldots\ldots(7),$$

$$\omega_x = F_x(lx + my + nz - c_x t), \text{ etc.} \ldots\ldots\ldots(8),$$

where f and F_x are arbitrary functions, restricting ourselves, for the present, to the consideration of plane waves†.

* Routh, *Rigid Dynamics*, Vol. II.

† The equation (7) states that Δ has the same value $f(d)$ over the two planes

$$lx + my + nz - c_1 t_1 = d,$$
$$lx + my + nz - c_1 t_2 = d;$$

therefore, if p_1, p_2 are the perpendiculars from the origin on these planes,

$$p_1 - p_2 = c_1(t_1 - t_2),$$

or the plane (l, m, n) travels parallel to itself with velocity c_1, every point of the plane being in the same state of dilatation. The plane $lx + my + nz =$ constant is thus the plane wave front, corresponding to the [dilatational] wave.

From (7), (8) we get, substituting in (5) and (6),

$$\rho c_1^2 = \lambda + 2\mu, \quad \rho c_x^2 = \mu.$$

We have thus two kinds of waves—a *pressural* wave propagated with velocity $\sqrt{\dfrac{\lambda + 2\mu}{\rho}}$ and a *distortional* wave, propagated uniformly, in all directions, with velocity $\sqrt{\dfrac{\mu}{\rho}}$.

76. If either $\Delta = 0$, or $\lambda + 2\mu = 0$ [or vanishingly small], we have only one kind of wave. Now $\Delta = 0$, when the medium is incompressible, i.e. when the coefficient of volume elasticity is large. Calling this coefficient k, we have $\lambda = \dfrac{3k\sigma}{1+\sigma}$, so that the vanishing of Δ corresponds to a large value of λ.

In any case, if we can assume with MacCullagh

$$\int W\, dxdydz = \int 2\mu\, (\omega_x^2 + \omega_y^2 + \omega_z^2)\, dxdydz \text{ ......(9)},$$

we get only one kind of wave. Stokes has pointed out* however that this expression for energy makes $X_y = -Y_x$†, so that the equilibrium condition is not satisfied or the medium is unstable.

This objection does not apply, if only the *total* strain energy has the value given by (9) and the energy is not localized in each element of volume or if the ether is 'labile' [Art. 96].

77. The conditions at the bounding surface are different according as $\lambda + 2\mu = 0$, or $\Delta = 0$ (in which latter case, λ is large).

In general, the conditions are (collecting surface integral

* *B. A. Report*, 1862.

† The expression for δW without the implication $X_y = Y_x$ is

$$X_x\, \delta e_{xx} + X_y\, \delta\left(\frac{\partial u}{\partial y}\right) + Y_x\, \delta\left(\frac{\partial v}{\partial x}\right) + \ldots;$$

$$\therefore \quad X_y = \frac{\partial W}{\partial u_y}, \quad Y_x = \frac{\partial W}{\partial v_x}, \text{ etc.,}$$

if we write $\quad u_y \equiv \dfrac{\partial u}{\partial y}, \quad v_x \equiv \dfrac{\partial v}{\partial x}$, etc.

Hence from (9) we get $X_y = -Y_x$.

terms [Arts. 73 and 74] and equating the coefficients of δu, δv, δw each to zero) that

$$(\lambda + 2\mu)\,\Delta l - 2\mu\,(m\omega_z - n\omega_y) - 2\mu\left[l\left(\frac{\partial v}{\partial y} + \frac{\partial w}{\partial z}\right) - m\frac{\partial v}{\partial x} - n\frac{\partial w}{\partial x}\right]$$

i.e.
$$\lambda\,\Delta l + 2\mu\left(\frac{\partial u}{\partial \nu} + m\omega_z - n\omega_y\right),\ \text{etc.,}$$

where
$$\frac{\partial u}{\partial \nu} = l\frac{\partial u}{\partial x} + m\frac{\partial u}{\partial y} + n\frac{\partial u}{\partial z},$$

should be continuous.

78. In particular, at the plane $x = 0$ $l = 1$, $m = 0$, $n = 0$.

$$\left.\begin{array}{l} \lambda\Delta + 2\mu\dfrac{\partial u}{\partial x} \\[2ex] \mu\left(\dfrac{\partial v}{\partial x} + \dfrac{\partial u}{\partial y}\right) \\[2ex] \mu\left(\dfrac{\partial w}{\partial x} + \dfrac{\partial u}{\partial z}\right) \end{array}\right\}\ \text{continuous.}$$

Besides these, there will be continuity of displacements at the surface. We shall take two special cases:

79. I. Let $u = 0$, $v = 0$, and let the axis of z be the line of intersection of the plane of the incident wave with the reflecting surface. Let also dashed letters refer to the second medium. Then

$$\lambda\frac{\partial w}{\partial z} = \lambda'\frac{\partial w'}{\partial z'} \quad \text{.....................(1),}$$

$$\mu\frac{\partial w}{\partial x} = \mu'\frac{\partial w'}{\partial x'} \quad \text{.....................(2),}$$

$$w = w' \quad \text{..........................(3).}$$

Note that (1) is identically satisfied, since w is independent of z.

80. II. Let $w = 0$, and u, v, independent of z. Then

$$\left.\begin{array}{l} \lambda\left(\dfrac{\partial u}{\partial x} + \dfrac{\partial v}{\partial y}\right) + 2\mu\dfrac{\partial u}{\partial x} \\[2ex] \mu\left(\dfrac{\partial v}{\partial x} + \dfrac{\partial u}{\partial y}\right) \\[2ex] u, \quad v \end{array}\right\}\ \text{continuous.}$$

Remembering that the plane of polarisation is defined as the plane through the ray and the normal to the reflecting surface, when light is polarised by reflection, case II is that of light polarised perpendicularly to the plane of incidence.

81. There are, thus, in case II, four conditions to determine two unknown quantities, viz., the intensities of reflected and refracted rays. There must, therefore, be two reflected and two refracted waves, even if the original wave was entirely transverse. But the energy of the incident wave is found* to be equal to the energy of the reflected and refracted (light or transverse) waves. Therefore no energy is absorbed by pressural waves, if they exist.

82. Green showed†, moreover, that the pressural wave will absorb very little energy, if $\frac{\lambda + 2\mu}{\mu}$ is either very large or very small. He assumed it to be very large, and on this understanding, and on the further assumption that the rigidity is the same in different media, showed that

(1) a difference of phase is introduced by reflection and refraction;

(2) the intensity of the reflected wave never vanishes.

As regards (1), the value of the difference of phase found does not agree with the result of experiment, while

as regards (2), the residual effect shown by his theory is much too great.

83. Proceeding now to the particular cases in which

$$\Delta = 0 \text{ or } \lambda + 2\mu = 0,$$

we observe that the conditions in case I are the same on either theory.

In case II, if $\Delta = 0$, the conditions are

$$\left.\begin{matrix} \mu \dfrac{\partial u}{\partial x} \\ \mu \left(\dfrac{\partial v}{\partial x} + \dfrac{\partial u}{\partial y}\right) \\ v \end{matrix}\right\} \text{ continuous,}$$

* Green, *Math. Papers.* † *Ibid.*

while, if $\lambda+2\mu=0$,

$$\left.\begin{matrix}\mu\dfrac{\partial v}{\partial y}\\ \mu\left(\dfrac{\partial v}{\partial x}+\dfrac{\partial u}{\partial y}\right)\\ v\end{matrix}\right\}, \text{ continuous;}$$

but since v must be the same and y the same, μ must be the same. The above conditions, thus, reduce to

$$\left.\begin{matrix}\dfrac{\partial v}{\partial x}+\dfrac{\partial u}{\partial y}\\ v\end{matrix}\right\}, \text{continuous.}$$

84. Case I. Consider the solution

$$w=A_1\,e^{ip(lx+my-Vt)}+A_2\,e^{ip(-lx+my-Vt)},$$
$$w'=A_3\,e^{ip'(l'x+m'y-V't)},$$

where $\left.\begin{matrix}l=-\cos i,\ l'=-\cos r\\ m=\sin i,\ m'=\sin r\end{matrix}\right\}$ (fig.).

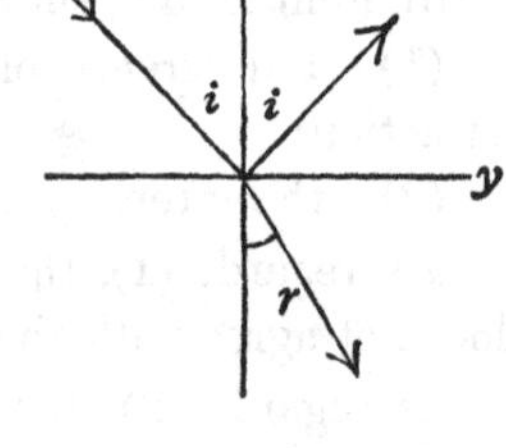

Now, since $w=w'$ at $x=0$,

$$pV=p'V', \text{ i.e. } \frac{V}{L}=\frac{V'}{L'},$$

where L, L' are the wave-lengths in the two media and $pm=p'm'$, i.e.

$$\frac{\sin i}{\sin r}=\frac{V}{V'}=\frac{L}{L'}.$$

Also, applying the conditions (2) and (3) of Art. 79,

$$A_2=-A_1\frac{\mu'\tan i-\mu\tan r}{\mu'\tan i+\mu\tan r},$$

while $$A_3=2\frac{A_1\,\mu\tan r}{\mu'\tan i+\mu\tan r}.$$

Again if E is the mean energy corresponding to

$$w=A\cos\frac{2\pi}{L}(Vt-x),$$

$$E=2\pi^2\frac{A^2V^2\rho}{L^2}=2\pi^2\frac{A^2\mu}{L^2}, \text{ since } \frac{\mu}{\rho}=V^2;$$

therefore, if E_1, E_2, E_3 be the energies of the incident, reflected, and refracted waves,

$$\frac{\sqrt{E_1}}{\frac{A_1}{L}} = \frac{\sqrt{E_2}}{\frac{A_2}{L}} = \frac{\sqrt{E_3}}{\frac{A_3}{L'}}.$$

Finally, if we take $\mu = \mu'$, we get

$$A_2 = A_1 \frac{\tan i - \tan r}{\tan i + \tan r},$$

$$A_3 = 2A_1 \frac{\tan r}{\tan i + \tan r}.$$

85. Case II. Light polarised perpendicularly to the plane of incidence. (Labile-ether theory.) Let

$$v_1 = A_1 \cos i \, e^{ip\,(x\cos i + y \sin i - Vt)}, \text{ (incident ray)},$$

$$v_2 = -A_2 \cos i \, e^{ip\,(-x\cos i + y \sin i - Vt)}, \text{ (reflected ray)},$$

$$v_3 = A_3 \cos r \, e^{ip'\,(x\cos r + y \sin r - V't)}, \text{ (refracted ray)},$$

and $$\frac{u_1}{\sin i} = \frac{v_1}{\cos i}; \quad \frac{u_2}{\sin i} = -\frac{v_2}{\cos i}; \quad \frac{u_3}{\sin r} = \frac{v_3}{\cos r}.$$

We have also $v_1 + v_2 = v_3$, at $x = 0$.

$$\therefore (A_1 - A_2) \cos i = A_3 \cos r.$$

This leads to a discontinuity of u. We shall see later that on the Electro-magnetic theory, electric displacement perpendicular to the surface is continuous, while along the surface the electric force alone is continuous. Continuity of u may be secured by introducing a pressural wave, propagated with zero velocity. Also, since $\frac{\partial u}{\partial y} + \frac{\partial v}{\partial x}$ is continuous,

$$(A_1 + A_2)\, p = A_3 p',$$

i.e. $$A_1 + A_2 = A_3 \frac{p'}{p} = A_3 \frac{\sin i}{\sin r}.$$

Thus $$(A_1 - A_2) \cos i = A_3 \cos r,$$

$$(A_1 + A_2) \sin r = A_3 \sin i;$$

$$\therefore \frac{A_1 - A_2}{A_1 + A_2} = \frac{\sin 2r}{\sin 2i},$$

whence $$\frac{A_1}{A_2} = \frac{\tan(i + r)}{\tan(i - r)},$$

and $$A_3 = A_1 \frac{2 \cos i \sin r}{\sin(i + r) \cos(i - r)}.$$

These are, also, Fresnel's formulae.

On this theory, the rigidity is the same in all media, but the density different in different media.

86. Summarising*, we find that we may hypothecate

(1) Null velocity of longitudinal waves ($\lambda + 2\mu = 0$).

(2) Incompressibility ($\Delta = 0$).

For (1), (*a*) The tangential components of displacement are continuous.

(*b*) The tangential stresses are continuous, while the normal stress is also continuous, if there is no loss of energy at reflection or refraction—i.e. in the case of the labile-ether of Kelvin. This also yields Fresnel's result.

For (2), (*a*) The three displacements are continuous, giving three conditions.

(*b*) Energy continuous, one condition.

(*c*) Tangential stresses continuous, two conditions (which are redundant).

SECTION II. THEORIES OF DOUBLE REFRACTION

87. The strain-energy function for a heterogeneous (æleotropic) solid contains six linear terms and twenty-one quadratic terms† in strains e_{xx}, etc.

If there are three planes of symmetry, then the terms involving the first power of shears cannot exist, since the form of this function should remain unchanged, if the signs of x, y, z are changed.

Therefore, the linear function will consist of three terms, viz., $Pe_{xx} + Qe_{yy} + Re_{zz}$ (say), and the quadratic function will contain $21 - 12 = 9$ constants. We may therefore write

$$2W = 2(Pe_{xx} + Qe_{yy} + Re_{zz}) + (A, B, C, F, G, H)(e_{xx}, e_{yy}, e_{zz})^2 + Le_{yz}^2 + Me_{zx}^2 + Ne_{xy}^2.$$

* Larmor, *Dynamical Theory of Electric and Luminiferous Media.*

† Six square terms and $\frac{6\cdot 5}{1\cdot 2}$ or 15 product terms.

If the body is originally unstressed, $P = Q = R = 0$; therefore, the equations of motion of the medium (of density ρ) will be

$$\rho \ddot{u} = \frac{\partial}{\partial x}\left(\frac{\partial W}{\partial e_{xx}}\right) + \frac{\partial}{\partial y}\left(\frac{\partial W}{\partial e_{xy}}\right) + \frac{\partial}{\partial z}\left(\frac{\partial W}{\partial e_{xz}}\right)$$

$$= \frac{\partial}{\partial x}\left[A\frac{\partial u}{\partial x} + H\frac{\partial v}{\partial y} + G\frac{\partial w}{\partial z}\right]$$

$$+ \frac{\partial}{\partial y} N\left(\frac{\partial u}{\partial y} + \frac{\partial v}{\partial x}\right) + \frac{\partial}{\partial z}\left[M\left(\frac{\partial u}{\partial z} + \frac{\partial w}{\partial x}\right)\right],$$

and two similar equations.

88. For an assumed mode of wave propagation in the form of a plane wave, let

$$u = \lambda\phi\,(lx + my + nz - Vt),$$
$$v = \mu\phi,$$
$$w = \nu\phi,$$

where ϕ is an arbitrary function. (λ, μ, ν) will then define the direction of displacement and $l^2 + m^2 + n^2 = 1$, while

$$(-\rho V^2 + a)\lambda + h\mu + g\nu = 0,$$
$$h\lambda + (-\rho V^2 + b)\mu + f\nu = 0,$$
$$g\lambda + f\mu + (-\rho V^2 + c)\nu = 0,$$

where $\quad Al^2 + Nm^2 + Mn^2 \equiv a, \quad (H + N)\,ml \equiv h,$ etc.

Hence $$\frac{a\lambda + h\mu + g\nu}{\lambda} = \ldots = \ldots,$$

showing that the directions of vibrations are along the principal axes of a certain quadric, the constants of which, however, depend on the direction of propagation itself.

89. If we assume, with Fresnel, that two of the vibrations are in the wave front, then,

$$l\lambda + m\mu + n\nu = 0 \quad \ldots\ldots\ldots\ldots\ldots\ldots\ldots(1).$$

Thus, we have

$$(al + hm + gn)\lambda + (hl + bm + fn)\mu + \ldots = 0.$$

Restoring the values of a, b, c, etc., in terms of A, B, C, we have

$$\lambda l\,[Al^2 + (H + 2N)\,m^2 + (G + 2M)\,n^2] + \ldots + \ldots = 0 \ \ldots(2).$$

And since (1) and (2) should be true for all values of λ, μ, ν, we must have

$$Al^2+(H+2N)m^2+(G+2M)n^2=K(l^2+m^2+n^2),$$

(where K is an undetermined constant), and two similar equations.

For this, it is necessary and sufficient that

$$A=B=C=K=H+2N=G+2M=F+2L.$$

On this proviso, we have

$$2W=K\Delta^2+L(e_{yz}{}^2-4e_{yy}e_{xx})+\ldots.$$

$$\therefore\ \delta\int 2W\,dx\,dy\,dz=\delta\int(K\Delta^2+4L\omega_x{}^2+4M\omega_y{}^2+4N\omega_z{}^2)\,dx\,dy\,dz.$$

[Arts. 73 and 74.] This leads to the equations

$$\rho\ddot{u}=K\frac{\partial\Delta}{\partial x}+2M\frac{\partial\omega_y}{\partial z}-2N\frac{\partial\omega_z}{\partial y},\ \text{etc.,}$$

i.e. $$\rho\ddot{\Delta}=K\nabla^2\Delta\ \text{(eliminating } \omega_x,\ \omega_y,\ \omega_z\text{)},$$

and $$\rho\ddot{\omega}_x=L\nabla^2\omega_x-\frac{\partial}{\partial x}\left[L\frac{\partial\omega_x}{\partial x}+M\frac{\partial\omega_y}{\partial y}+N\frac{\partial\omega_z}{\partial z}\right],\ \text{etc.}$$

(eliminating Δ) [Art. 74].

90. Putting now $2\omega_x=\lambda'\phi(lx+my+nz-Vt)$, we have $(\rho V^2-L)\lambda'=-(Ll\lambda'+\ldots+\ldots)l$, etc., which, on account of the condition $l\lambda'+m\mu'+n\nu'=0$*, yields

$$\frac{l^2}{\rho V^2-L}+\ldots+\ldots=0.$$

This, of course, gives Fresnel's wave surface.

We have, however,

$$\rho V^2=L\lambda'^2+M\mu'^2+N\nu'^2,$$

i.e. $$V^2=a^2\lambda'^2+b^2\mu'^2+c^2\nu'^2,$$

if a, b, c are the principal wave velocities; but, on Fresnel's hypothesis, $V^2=a^2\lambda^2+b^2\mu^2+c^2\nu^2$ [Art. 54], which shows that λ', μ', ν', defining the axis of rotation on the present theory, coincide with λ, μ, ν, defining the direction of Fresnel's displacement. The axis is therefore perpendicular to the plane of polarisation; that is, the direction of vibrations on Green's theory is *in* the plane of polarisation.

* Since rotations and displacements are necessarily at right angles to each other, this also follows from the relation

$$2\omega_x=\frac{\partial w}{\partial y}-\frac{\partial v}{\partial z}=(\nu m-\mu n)\phi$$

from (1) of Art. 89.

91. In order to obviate this difficulty, Green included the terms involving P, Q, R along with second order terms for e_{xx}, etc. in the form

$$e_{xx}=\frac{\partial u}{\partial x}+\frac{1}{2}\left[\left(\frac{\partial u}{\partial x}\right)^2+\left(\frac{\partial v}{\partial x}\right)^2+\left(\frac{\partial w}{\partial x}\right)^2\right].$$

He, further, supposed that for waves perpendicular to any two of the principal axes and propagated by vibrations in the direction of the third axis, the velocity of propagation is the same. This forced relation enabled him to obtain Fresnel's wave surface as well as the direction of vibration given by Fresnel's theory.

Moreover, he took density to be the same in every direction but rigidity different in different directions.

92. In applying the theory to the case of crystalline refraction it is found necessary to suppose further forced relations between the rigidity in different media, for which there does not seem to be any justification.

93. MacCullagh's theory.

We have seen [Art. 76] that the strain-energy function in the case of an isotropic solid can be written (if $(\lambda+2\mu)\Delta=0$)

$$=2\mu\left(\omega_x^2+\omega_y^2+\omega_z^2\right),$$

since the other terms yield only surface integrals, when the variation of the function is taken (even if we do not assume the ether to be labile), and that MacCullagh assumed this to be the actual form of the function. Generalizing this, for a crystal, he assumed that

$$W=2\left(L\omega_x^2+M\omega_y^2+N\omega_z^2\right),$$

while the density was taken uniform and the same in all media.

The equation of motion of the disturbed medium accordingly becomes

$$\rho\ddot{u}=2M\frac{\partial\omega_y}{\partial z}-2N\frac{\partial\omega_z}{\partial y}\text{, etc.}$$

or

$$\rho\ddot{\omega}_x=L\nabla^2\omega_x-\frac{\partial}{\partial x}\left[L\frac{\partial\omega_x}{\partial x}+M\frac{\partial\omega_y}{\partial y}+N\frac{\partial\omega_z}{\partial z}\right]\text{, etc.}$$

These equations of torsional vibrations are of course the same as Green's. [Art. 89.]

On this theory, however,

$$\frac{\partial u}{\partial x}+\frac{\partial v}{\partial y}+\frac{\partial w}{\partial z}=0,$$

so that there is no compression of the medium involved, whether the ether is assumed to be compressible or not. Otherwise, we have the same difficulty as on Green's theory.

94. A hypothesis that the density is the same in all media gives correct results, but this leads to the conclusion that in an isotropic medium, there are two polarising angles*.

95. The general import of all this analysis may be thus summed up.

Assuming that the ethereal medium is of the nature of an elastic solid, it is easy to calculate (as was done by Green) the effect of a small disturbance (in the nature of an elastic displacement). It is, then, found that two waves will be set up, a pressural wave and a torsional wave, the latter of which alone can constitute that of light. Green, accordingly, supposed the pressural wave to be propagated with very great (infinite) velocity. But such a supposition proves to be unacceptable, inasmuch as it is incapable of explaining the well-known phenomenon of polarising angles. In the next place, in explaining double refraction on the basis—naturally—of the theory of crystalline—æleotropic—elasticity with three planes of symmetry, he assumed that in a crystal, ethereal rigidity is different in different directions, but the (ethereal) density the same throughout. This led to the erroneous conclusion that there are two polarising angles. His theory, moreover, led to the further conclusion that vibrations of polarised light are parallel to the plane of polarisation, in contradistinction to Fresnel's, according to which the vibrations are perpendicular to this plane.

It was, therefore, natural to inquire if the velocity of propagation of the pressural wave might not be zero consistently with the principle of energy, and Lord Kelvin showed that this is allowable, provided we postulate certain properties (e.g. that of null velocity) of the ether at the bounding surface, which he

* Rayleigh, *loc. cit.*

called 'labile.' MacCullagh had previously argued that, inasmuch as light disturbance was transversal, the strain-energy function for the ethereal medium, so far as its optical properties are concerned, must be a function (quadratic) of its molecular rotations only, and this yields the same result as Lord Kelvin's theory.

96. It follows that all the optical theories which have been at all successful in giving a fairly satisfactory account of the mechanism of light propagation depend on the assumption that the potential energy of light disturbance in a homogeneous medium is of the form (except for a constant factor)

$$\iiint (\omega_x^2 + \omega_y^2 + \omega_z^2)\, dx\, dy\, dz,$$

in order that it shall account for the non-existence of a pressural wave.

Thus, on MacCullagh's theory, ω_x, ω_y, ω_z are the molecular rotations of the elastic medium, behaving as an elastic solid.

On the labile-ether theory also, the strain-energy function is found to assume the same form—together with certain surface integrals which vanish on account of the 'labile' properties of the medium at the boundaries.

[On the electromagnetic theory, ϖ_x, ϖ_y, ϖ_z are the electric displacements of Maxwell, Art. 190.]

It will be seen, however, that if the above expression is to be true for each element of volume, the equilibrium condition (as was pointed out by Stokes) is not satisfied, for it makes $X_y = -Y_x$, where X_y, Y_x are the mutually perpendicular tangential stresses on the opposite faces of an elementary parallelepiped. Now, this difficulty can be met, if we suppose that the above energy is not localized in each element of volume. But, in that case, we must postulate the existence of certain terms in the energy function for each such element, which should disappear when integrated throughout the whole volume of the medium. That is, we must admit the existence of a certain amount of intrinsic energy in the medium which takes part in the propagation of light. This is virtually the view adopted by Larmor who has pointed out that if we could postulate some concealed (rotational) phenomenon going on in each element, the kinetic reaction of which can give rise to the requisite couple, the instability pointed out by Stokes may be provided for. He goes on to remark that the explanation of gravitation is still outstanding and necessitates some structure or property quite different from or probably more fundamental than simple rotational elasticity of the ether and simple molar elasticity of material aggregates in it, and this property may be operative in the manner here required.

No doubt, the 'labile-ether' theory does not require this saving hypothesis, as it only postulates a certain form of energy function for a whole volume, not localized in each element (together with certain surface integrals, which must be supposed to vanish). At the same time, although, on this mode of regarding the question, the analytical difficulty is met, the hypothesis of 'labile' ether leaves the peculiar property of the medium, somewhat ill-defined, while, on the other hand, if the existence of intrinsic energy were admitted, we should have a clearer view of the intimate processes, whose resultant alone is exhibited by the final analytical result postulated.

It is remarkable that Fermat's law leads to similar conclusions.

As regards the energy function on the electromagnetic theory, the formal similarity of the expression for potential energy of the rotationally elastic medium of MacCullagh and the electrostatic energy enables us to interpret the intimate nature of the electric field in the same way, that is, on the basis of the same postulate of a certain amount of intrinsic energy of the ethereal medium.

97. On the whole, therefore, neither the phenomena of reflection and refraction nor that of double refraction can be explained on a pure elastic solid theory. Some action between ether and matter must be postulated, even to explain these, the simplest optical phenomena. As for those which, like dispersion, aberration, and fluorescence, obviously depend on this interaction, this postulate seems to be essential. It should be noted, however, that in postulating æleotropy in regard to the ether, this action has in a manner been taken account of—though not, perhaps, in the way in which it is actually operative.

98. Cauchy was the first to attempt to construct a theory on these lines. Ether atoms in a body, such as a crystal, he supposed disposed as it were in shells, round the matter atom, in such a manner as to have different elastic properties at different points of the same shell. The shells, moreover, are, according to him, regularly placed and the properties of the ether repeat themselves at similar points in the different shells. Hence, the constant in the equations of motion is found to be a periodic function of the equilibrium position of the molecules, and for optical effects we have to do with the average displacement over a small volume of the medium.

But he still obtained the pressural wave and in order to account for its non-existence supposed $\lambda + 2\mu$ to be negative and hence the velocity of propagation of such a wave, imaginary or non-existent. Cauchy's attempt must therefore be held to have been wholly unsuccessful*.

SECTION III. MODIFIED ELASTIC SOLID THEORY

99. The next notable attempt in this direction was that of Lord Rayleigh. The analogy of a solid moving in a fluid would suggest that the first effect of a matter molecule in a transparent body would be to alter the apparent density of the solid, and, conceivably, this alteration might depend in a body like a crystal on the direction of vibration. If this is admitted, and further, if it is allowable to suppose that the rigidity of the elastic ether is unaltered by the presence of matter, we get a wave surface in a biaxial crystal, which is Fresnel's wave surface, provided the ether is assumed to be 'labile.'

The above considerations led Lord Rayleigh to adopt for the kinetic energy T the value given by

$$2T = \int (\rho_x \dot{u}^2 + \rho_y \dot{v}^2 + \rho_z \dot{w}^2)\, dx\, dy\, dz.$$

* Glazebrook, *B. A. Report*, 1885.

If further the elastic property of the ether is unaltered by the presence of matter, we shall have

$$2W = (\lambda + 2\mu)\Delta^2 + 4\mu[(e_{xy}^2 - 4e_{xx}e_{yy}) + \ldots + \ldots];$$

therefore the equations of motion are

$$\rho_x \ddot{u} = (\lambda + \mu)\frac{\partial \Delta}{\partial x} + \mu \nabla^2 u, \text{ etc. [Art. 74.]}$$

Assuming $u = l'\phi(lx + my + nz - Vt)$, etc.,

we have $\rho_x V^2 l' = (\lambda + \mu) l (ll' + mm' + nn') + \mu l'$, etc.

$$\therefore \frac{l^2}{\rho_x V^2 - \mu} + \frac{m^2}{\rho_y V^2 - \mu} + \frac{n^2}{\rho_z V^2 - \mu} = \frac{1}{\lambda + \mu}.$$

100. If the ether is labile, we must put $\lambda + 2\mu = 0$,

i.e. $$\frac{l^2}{\rho_x V^2 - \mu} + \ldots + \ldots + \frac{l^2 + m^2 + n^2}{\mu} = 0;$$

$$\therefore \frac{l^2}{V^2 - a^2} + \ldots + \ldots = 0, \text{ where } \frac{\mu}{\rho_x} = a^2, \text{ etc.,}$$

which gives Fresnel's wave surface.

If we put $\dfrac{1}{\lambda + \mu} = 0$ (that is, assume the ether to be incompressible), we get

$$\frac{l^2}{\rho_x V^2 - \mu} + \ldots + \ldots = 0, \text{ i.e. } \frac{l^2 a^2}{V^2 - a^2} + \ldots + \ldots = 0 \quad \ldots(1).$$

Comparing this with the equation of Fresnel's wave surface,

$$\frac{a^2 x^2}{r^2 - a^2} + \ldots + \ldots = 0 \quad \ldots\ldots\ldots\ldots\ldots\ldots(2),$$

we notice that since in (1) V is equal to the perpendicular on the tangent plane and (l, m, n) is the normal to the tangent plane, (2) is the pedal of (1).

Fresnel's wave surface is, therefore, the pedal of the wave surface obtained on the theory of an incompressible ether supposed to be elastically isotropic but of æleotropic inertia.

101. Again, we have, if $\lambda + 2\mu = 0$,

$$\left(\frac{V^2}{a^2} - 1\right) l' = - l (ll' + mm' + nn'), \text{ etc.} \quad \ldots\ldots\ldots(3).$$

Multiplying the equations in order by l, m, n and adding, we get

$$\frac{ll'}{a^2}+\frac{mm'}{b^2}+\frac{nn'}{c^2}=0.$$

Hence (multiplying the equations in order by $\frac{l}{a^2}, \frac{m}{b^2}, \frac{n}{c^2}$, and adding)

$$V^2\left(\frac{l'^2}{a^4}+\ldots+\ldots\right)=\left(\frac{l'^2}{a^2}+\ldots+\ldots\right) \quad \ldots\ldots\ldots(4).$$

Therefore if we write

$$\frac{L}{\frac{l'}{a^2}}=\ldots=\frac{Ll'+\ldots+\ldots}{\frac{l'^2}{a^2}+\ldots+\ldots}=\frac{1}{\left(\frac{l'^2}{a^4}+\ldots+\ldots\right)^{\frac{1}{2}}} \quad \ldots(5),$$

we get $$V^2=a^2L^2+b^2M^2+c^2N^2,$$

showing that L, M, N define the direction of the displacement, on Fresnel's theory.

Now, from (3) if $ll'+\ldots+\ldots=\cos\psi$,

we get $$V^2\left(\frac{l'^2}{a^2}+\ldots+\ldots\right)=\sin^2\psi;$$

therefore, from (4)

$$\frac{\left(\frac{l'^2}{a^2}+\ldots+\ldots\right)^2}{\left(\frac{l'^2}{a^4}+\ldots+\ldots\right)}=\sin^2\psi=(Ll'+Mm'+Nn')^2 \text{ from (5).}$$

Therefore the inclination of l', m', n' to l, m, n is the complement of the inclination of l', m', n' to L, M, N; that is

If OP is the wave normal, OR the direction of the ray, PR the direction of the displacement, on Fresnel's theory then, PN is the direction of the displacement on the present theory where PN is perpendicular to OR.

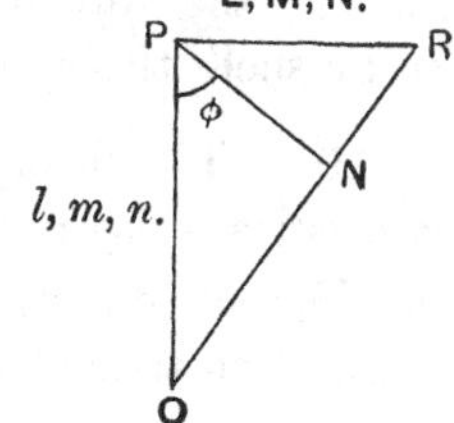

Again, since

$$\left(\frac{V^2}{a^2}-1\right)l'=-l\,(ll'+mm'+nn'),$$

we have $$\frac{l'}{\frac{a^2l}{V^2-a^2}}=\ldots=K \text{ (say)} \ldots(6).$$

But for Fresnel's wave surface, if (x, y, z) is the point of contact of the tangent plane (l, m, n) with the surface,

$$x = lV\frac{r^2 - a^2}{V^2 - a^2} = V\frac{l'}{Ka^2}(r^2 - a^2); \quad \text{........(7)}$$

$$\therefore \quad \frac{a^2x^2}{r^2 - a^2} = \frac{V}{K}l'x,$$

$$\text{or} \quad l'x + m'y + n'z = 0,$$

since $$\frac{a^2x^2}{r^2 - a^2} + \ldots + \ldots = 0,$$

that is, (l', m', n') is perpendicular to the ray.

Let O be the centre of the wave surface, OP the normal, OR the ray, RP the tangent line and PN perpendicular to OR.

Then by the triangle of velocities

$$\overline{NP} + \overline{PO} + \overline{ON} = 0.$$

Hence, by projection,

$$NP \,.\, L - Vl + \frac{V^2}{r} \,.\, \frac{x}{r} = 0 \quad \text{.................(8);}$$

since $OP = V$, $OR = r$ $\left(\text{and, therefore, } ON = \frac{V^2}{r}\right)$; also, L, M, N are the direction-cosines of PR.

$$\therefore \text{ from (8), } NP \,.\, L = lV - V^2\frac{x}{r^2}$$

$$= lV\left(1 - \frac{V^2}{r^2} \cdot \frac{r^2 - a^2}{V^2 - a^2}\right) \text{ from (7),}$$

$$= \frac{lV}{r^2}\left[a^2\frac{V^2 - r^2}{V^2 - a^2}\right];$$

$$\therefore \quad \frac{L}{l'} = \frac{M}{m'} = \frac{N}{n'}, \text{ from (6);}$$

that is, the direction of vibration is in the plane PON as on Fresnel's theory but not in the wave front*.

102. The theory is based on the hypothesis that the ether is elastically isotropic but that its optical density is different in different directions and different in different media. In attempting to analyse this hypothesis, we observe that from (4)

* Glazebrook, *Phil. Mag.* 1888.

of Art. 74 the equation of motion of the ether may be written, on the theory of contractile (or labile) ether, in the form

$$\rho \ddot{u} = \mu \nabla^2 u + X \text{ ....(1)},$$

on the assumption that X is the force exerted by matter molecules.

If ρ_1 is the density of matter and (U, V, W) its displacement, then the equation (1) becomes

$$\rho \ddot{u} = \mu \nabla^2 u - \rho_1 \ddot{U} \text{ ....(2)},$$

for X may obviously be assumed to be of the form

$$-\rho_1 \ddot{U},$$

while U, the displacement of the matter molecule, may be taken to have the form given by

$$U = f(u),$$

the nature of f depending on the nature of the crystal*.

103. Thus, if we take $U = Au$, etc. in an isotropic medium, we have to replace ρ by $\rho + \rho_1 A$, and the equation (2) of Art. 102 may be written

$$(\rho + \rho_1 A)\, \ddot{u} = \mu \nabla^2 u.$$

In the case of æleotropy of inertia, if we write

$$U = A_1 u, \quad V = A_2 v, \text{ etc.},$$

we get the expression for the kinetic energy, in the form

$$\tfrac{1}{2} \int (\rho_x \dot{u}^2 + \rho_y \dot{v}^2 + \rho_z \dot{w}^2)\, dx\, dy\, dz.$$

Again, in the case of media exhibiting rotatory polarisation, a form that will give the experimental result is

$$U = Au + B\left(\frac{\partial v}{\partial z} - \frac{\partial w}{\partial y}\right) \equiv Au - 2B\varpi_x, \text{ etc.}$$

In order to explain dispersion we may take the relation

$$U = Au - D\nabla^2 u, \text{ etc.},$$

and finally aberration may be supposed to be due to the form of X given by

$$-\rho_1 \left(\frac{\partial}{\partial t} + v_x \frac{\partial}{\partial x} + v_y \frac{\partial}{\partial y} + v_z \frac{\partial}{\partial z}\right)^2 u,$$

where v_x, etc. are the velocities of the matter molecules.

Or we may assume the form for X (in the case of isotropy) to be

$$-n\, \frac{d^2 (u - U)}{dt^2},$$

* Boussinesq, *loc. cit.*

and remembering that U must necessarily be small compared with u, if there is no absorption of energy, we get the equation

$$(\rho + n)\,\ddot{u} = \mu\nabla^2 u,$$

and admitting n to be different for different directions, we get the equations appropriate to æleotropy of inertia.

For dispersion, if we adopt the view that it arises from absorption of energy, due to the near synchronism of the ether period with the period of matter molecules, the force on the matter molecules may be taken to be of the form

$$\rho' \frac{d^2}{dt^2}(u - U) - \alpha^2 U - \gamma^2 \dot{U},$$

where α, γ are constants.

Thus, for the ether we have

$$\rho\ddot{u} = \mu\nabla^2 u - \rho' \frac{d^2}{dt^2}(u - U),$$

and for matter

$$\rho_1 \ddot{U} = \rho' \frac{d^2}{dt^2}(u - U) - \alpha^2 U - \gamma^2 \dot{U},$$

where ρ_1 is the density of the matter molecules actually taking up the vibrations*.

104. The above mode of writing down differential equations in order to obtain a particular result might at first sight appear to recall Fresnel's mode of explaining double refraction; but it cannot be denied that it has what may be described as a dynamical substratum, a fuller view of which can only be obtained on a clearer apprehension of the intimate nature of the interaction between matter and ether, which these equations serve provisionally to represent. This fuller view, as we shall see, is partly supplied by the electron theory [Ch. IV].

* Glazebrook, *loc. cit.*

CHAPTER III

ELECTRO-MAGNETIC THEORY

Section I. Statical and Dynamical Electricity

105. We have seen that an attempt to construct the properties of the ether from the properties of light, on the assumption that the ether behaves as an elastic solid when it takes part in the propagation of light, has been so far only partially successful. For, on any of the theories that have at all succeeded in giving an approximately satisfactory account of optical phenomena, the question is left entirely doubtful, as to how the ethereal density and the ethereal rigidity arise and what is the intimate nature of the forces that the interaction of matter and ether brings into play. Some light may be thrown on these points if the electro-magnetic field should turn out to be, as we have ample justification for holding that it is, the ethereal medium of optical theory.

"Although," to quote Maxwell, "to fill all space with a new medium, whenever any new phenomenon is to be explained, is by no means philosophical, if the study of two different branches has independently suggested the idea of a medium and if the properties which must be attributed to the medium in order to account for electro-magnetic phenomena are of the same kind as those which we attribute to the luminiferous medium, in order to account for the phenomena of light, the evidence for the physical existence of the medium will be considerably strengthened."

106. Now, the ethereal medium, for the purposes of the propagation of light, on the theory of undulation, must be a receptacle of energy The first question that must therefore be considered is, Is the electro-magnetic medium also a receptacle of energy?

Faraday answered this question in the affirmative and postulated a state of strain in the electrostatic and electro-magnetic field, imaged by lines of electric and magnetic forces. To him, all electric and magnetic phenomena reduced themselves to changes in the number and distribution of lines of force, the existence of which were, according to him, effectively demonstrated by the phenomena of static and electro-magnetic induction. Maxwell studied these properties mathematically and in doing so showed that an electro-magnetic disturbance should be propagated with the velocity of light. This we proceed to consider.

107. The well-known gravitational law of force is expressed, as we know, as that of an action at a distance. And, of course, the law of electric action can also be expressed in the same way, viz., that the repulsion between two point charges e, e' is $\frac{ee'}{Kr^2}$, where r is the distance between the charges and K the specific inductive capacity of the medium in which the action takes place.

108. Now from this known law of electric action in any medium and without any assumption as to the intimate nature of electric phenomena, it is easy to deduce that if a charge e is placed in any medium of specific inductive capacity K,

V = the potential at any point of the medium, defined in the same way as gravitational potential,

dS = an element of surface S at that point, dn an element of the normal*,

F_n = the electric intensity at that point, resolved normally to the surface, and

$d\varpi$ = solid angle subtended at that point by the surface S, then

$$-\int K\frac{\partial V}{\partial n}dS = \int KF_n dS = \int K\frac{e}{Kr^2}dS\cos\theta = e\int d\varpi = e\varpi,$$

θ being the angle between r and the normal to dS.

And if the surface completely encloses a charge e,

$$\int KF_n dS = 4\pi e.$$

* Outward-drawn (in all cases).

Let N_0 and N be two numbers such that (by a choice of units) we can put $4\pi e = N_0$, $\frac{\varpi}{4\pi} = \frac{N}{N_0}$, then

$$\int K F_n dS = N,$$

the integration being taken over S.

The equation can be stated in words, thus:

The normal induction over S = number of tubes of induction intercepted by S, provided we define

(1) $\int KF_n dS$ over S as the normal induction over S,

(2) a unit tube to be the cone generated by lines of force issuing from the charge such that its solid angle $= e^{-1}$.

When all the tubes are drawn round the point-charge, the field is obviously mapped out, so that electric action, due to the charge can be exhibited by means of these lines.

109. In order to exhibit the effects due to a charged conductor, which is known from well-known experiments to be an equipotential surface, we have to draw lines of force (normal to the charged conductor) along the periphery of an element dS of the surface of the conductor.

Consider the tubular surface generated by these lines and limited by consecutive equipotential surfaces on either side of the conductor:

The charge enclosed by the surface is σdS, if σ is the surface charge.

Applying the theorem $\int KF_n dS = 4\pi e$, we have

$$(KF_n)_1\, dS_1 + (KF_n)_2\, dS_2 = 4\pi\sigma dS,$$

$(KF_n)_1\, dS_1$ referring to the inner equipotential, and $(KF_n)_2\, dS_2$ to the outer.

But $(F_n)_1 = 0$ since the potential inside a hollow conductor (which does not enclose a charge) is constant;

$$\therefore\ (KF_n)_2 = 4\pi\sigma, \text{ since } dS_2 = dS, \text{ ultimately.}$$

If now $(KF_n)_2\, dS = dN$, then $4\pi\sigma dS = dN$.

That is, the charge on the element dS is proportional to the number of tubes of force issuing from the element.

By drawing tubes of force, therefore, we can represent the electrical condition of the conductor.

110. If we fix our attention on the electrified conductors only, these tubes of force are to be regarded as serving merely to represent geometrically the state of the conductors. But since N depends on K, we are naturally led to conclude that electrical action must depend on the nature of the medium—at any rate to some extent.

These tubes issuing from a charged conductor must end somewhere. The locus of their extremities must be another conductor (or another part of the same conductor) which therefore presents to the tubes an opposite aspect to that which is presented by the conductor or portion of the conductor from which they issue. The second conductor is said to be oppositely charged. Thus a charge of one kind means the existence of an equal and opposite charge. This is also a fundamental fact of experiment, viz., that equal and opposite charges always subsist together; the charge of the conductor from which a tube issues being called (provisionally) positive, the charge of the conductor at which they end must be called negative.

111. Consider now a condenser consisting of two infinite parallel conducting plates at potentials V_1 and V_2.

If the medium is air, the number of tubes of force will be proportional to $\frac{V_1 - V_2}{d}$, where d is the distance between the plates.

If it is a medium of specific inductive capacity K, this number is proportional to $K\frac{V_1 - V_2}{d}$; that is, if $K > 1$, there will be a greater aggregation of these tubes for the same electromotive intensity and a correspondingly greater accumulation of charge, positive and negative, on the two plates.

112. We may, therefore, describe the electric state of a condenser, as being due to a certain modification—a certain mechanical condition or constraint of the medium, the conductors being boundaries of the medium, whose electric conditions define the boundary conditions of the medium, as in the theory of elasticity.

113. From what we have seen, every case of electrification is that of a condenser, the electrified body forming one boundary, and the oppositely electrified body, the opposite boundary of the medium in which these are situated.

As this view of electric phenomena can be justified on experimental grounds, it follows that

(1) the electric energy is the energy due to this peculiar condition of the medium, and

(2) the tubes of force are to be viewed as being related in some manner to this mechanical—this strained—condition, so that they must be regarded as having a physical and not merely a geometrical significance.

114. We know that along a tube of force spanning between surface charges σ, $-\sigma$ (dS being the elementary cross-section of a tube and F_n = electric intensity along it)

$$KF_n dS = 4\pi\sigma dS;$$

$$\therefore\ KF_n^2 dS dn = -4\pi\sigma \frac{\partial V}{\partial n} dS dn \quad \ldots\ldots\ldots\ldots\ldots(1),$$

where $-\dfrac{\partial V}{\partial n}$ is the electric intensity, at any point, along a tube of force and dn an element of a line of force. If, now, $d\tau$ is an element of volume of the medium, we have $d\tau = dS \,.\, dn$ and we get from (1),

$$\int K\frac{F_n^2}{8\pi}d\tau = -\frac{1}{2}\int \sigma dS dV = \frac{1}{2}\int \sigma dS(V_0 - V_1) \ \ldots(2),$$

since σdS is constant along such a tube, where

V_0 = potential at the charge σ,

V_1 = potential at the charge $-\sigma$.

But, if W is the electrostatic energy of the system,

$$\int\sigma(V_0 - V_1)\, dS = 2W; \quad \therefore\ W = \int\frac{K}{8\pi}F_n^2 d\tau.$$

115. The electrostatic energy of the system is therefore accounted for by conceiving a distribution of energy throughout the volume of the dielectric of amount $\dfrac{K}{8\pi}F_n^2$ per unit volume. In fact* "the equality of these two expressions [(2) of Art 114] may be interpreted physically, by conceiving the

* Maxwell, *Elect. and Mag.* Vol. I.

physical relation between the electrified bodies, either as the result of the state of the intervening medium, or as a result of a direct action between the electrified bodies at a distance. If we adopt the latter conception, we may determine the laws of action, but we can go no further in speculating on its cause. If, on the other hand, we adopt the conception of an action through a medium, we are led to inquire into the nature of that action in each part of the medium."

"If we now proceed to investigate the mechanical state of the medium on the hypothesis that the mechanical action between electrified bodies is exerted by and through the medium, we find that the medium must be in a state of mechanical stress."

116. Since, on this view, a conductor is the boundary of a dielectric, the electrification observed at the conductor may be assumed to exist at each equipotential surface; but since at a metallic conductor, the surface density of electricity $= \frac{K}{4\pi} F_n$, we may conceive at each point of a homogeneous isotropic medium, a surface charge σ (+, −) distributed over each element of an equipotential, where

$$\sigma = \frac{K}{4\pi} F,$$

F being the electric intensity at the point. This amounts to the assumption that the medium is *polarised* (Art. 130) as the result of the action of F. σ has been called by Maxwell the electric displacement at the point.

117. In the same way, since the electrostatic pressure at the surface of a conductor is $\frac{2\pi\sigma^2}{K}$ directed away from the conductor into the medium of dielectric constant K, we may account for this condition at the boundary, if we admit that there is a stress along a tube of force of the nature of a tension of this magnitude, at every point.

If this is T, it follows that $T = \frac{2\pi\sigma^2}{K} = \frac{1}{8\pi} KF^2$.

118. Let us next assume that the principal axes of stress at any point of the medium are in the direction of the line of force through this point and any two directions at right angles to each other, in the plane perpendicular to this line.

Describe an elementary tube enclosing the point in the form of a rectangular parallelepiped, two of whose faces are perpendicular to a line of force through the point. Let the volume of the elementary tube be $d\tau$, and let $P =$ the stress across the faces which are parallel to the line of force. Then the energy arising from T, P, P at the element of volume $d\tau$

$$= \int (-T - 2P)\, d\tau.$$

But this must be

$$= \frac{1}{8\pi} \int K F^2 d\tau.$$

We have seen, however, that $T = \dfrac{K}{8\pi} F^2$ (on a certain hypothesis);

$$\therefore \quad P = -\frac{K}{8\pi} F^2$$

(which is, therefore, of the nature of a pressure).

119. The various assumptions that we have made in arriving at the above result indicate the limitations to which it is subject. We arrive, in fact, at a particular distribution of stress which is consistent with known surface conditions and known volume integral of energy. This corresponds to $W = \psi_1$ [Art. 69].

120. To investigate generally the stresses that must produce this strained condition of the medium, we may proceed as follows:

If X_x, X_y, etc. are the stress components in three directions at right angles to each other at any point as in an elastic medium, under body forces ρX, etc. per unit volume, we must have

$$\frac{\partial X_x}{\partial x} + \frac{\partial X_y}{\partial y} + \frac{\partial X_z}{\partial z} + \rho X = 0, \text{ etc.}$$

Maxwell takes $X = -\dfrac{\partial V}{\partial x}$, V being the potential at the point due to the charge to which the supposed stresses are due, and ρ to be the volume density of this charge, supposed to be distributed throughout the medium as free electricity; i.e. the volume density of charge associated with electric displacement due to E.M.F. impressed on the field. ρX, etc. are thus identified with the body forces of the ethereal medium, the density of electric charge being assumed to be the same as its mass per unit volume.

121. A particular set of solutions is that given by Maxwell:

Since
$$\frac{\partial X_x}{\partial x}+\frac{\partial X_y}{\partial y}+\frac{\partial X_z}{\partial z}=\rho X,$$
and
$$\nabla^2 V+4\pi\rho=0,$$
we have
$$\frac{\partial X_x}{\partial x}+\frac{\partial X_y}{\partial y}+\frac{\partial X_z}{\partial z}=-\frac{1}{4\pi}\frac{\partial V}{\partial x}\left[\frac{\partial^2 V}{\partial x^2}+\frac{\partial^2 V}{\partial y^2}+\frac{\partial^2 V}{\partial z^2}\right],$$
so that it is easy to verify that
$$X_x=\frac{1}{8\pi}\left[\left(\frac{\partial V}{\partial x}\right)^2-\left(\frac{\partial V}{\partial y}\right)^2-\left(\frac{\partial V}{\partial z}\right)^2\right],$$
$$X_y=\frac{1}{4\pi}\frac{\partial V}{\partial x}\cdot\frac{\partial V}{\partial y},\ \text{etc.}$$
will satisfy the equations.

To obtain the general solution, we have to add to the above, the solution obtained by making
$$\frac{\partial X_x}{\partial x}+\ldots+\ldots=0.$$

The stresses have further to satisfy surface conditions at the boundaries of the dielectric, that is at the surface of the conductors.

The corresponding strains have also to satisfy the equations of compatibility of Saint-Venant*. [Art. 123.]

122. It is clear that the above system of stresses is equivalent to that already obtained.

For, taking the axes as in Art. 118, we have
$$\frac{\partial V}{\partial x}=-F,\quad \frac{\partial V}{\partial y}=0=\frac{\partial V}{\partial z};$$
$$\therefore\quad X_x=\frac{KF^2}{8\pi},\quad X_y=0=X_z=Y_z,\quad Y_y=-\frac{KF^2}{8\pi}=Z_z,$$
so that this system of stresses involves the assumptions in Arts. 116–121.

* Love's *Elasticity*, *loc. cit.*

It will be seen, moreover, that the system is consistent with the boundary condition, namely that the normal stress at the boundary is a tension of amount $\frac{KF^2}{8\pi}$. [See Appendix II.]

123. The above system of stresses is however inconsistent with the condition of stability, if we assume that the properties of the medium which is supposed to be strained are consistent with the ordinary laws of elasticity, satisfying the conditions of compatibility enunciated by Saint-Venant.

For this, let e_{xx}, e_{xy}, etc. be the strains at any point. Then it is easy to show that the above system of stresses gives rise to strains given by

$$e_{xx} = \frac{K}{8\pi E}(1+2\sigma) F^2,$$

where K is the specific inductive capacity, E, Young's modulus, σ, Poisson's ratio, and

$$e_{xy} = 0, \text{ etc.},$$

$$e_{yy} = -\frac{F^2 K}{8\pi E}, \text{ etc.}$$

Applying the compatibility equations, we have

$$\left(\frac{\partial^2}{\partial y^2} + \frac{\partial^2}{\partial z^2}\right) F^2 = 0,$$

$$\left(-\frac{\partial^2}{\partial x^2} + (1+2\sigma)\frac{\partial^2}{\partial z^2}\right) F^2 = 0,$$

$$\left(-\frac{\partial^2}{\partial x^2}(1+2\sigma) - \frac{\partial^2}{\partial z^2}\right) F^2 = 0,$$

$$\frac{\partial^2 F}{\partial y \partial z} = 0, \text{ etc.}$$

Hence $F^2 = ax + by + cz$, which is, obviously, impossible.

If, however, the phenomena are not statical (and they are not if $W = \psi_1$ [Art. 119]), the above argument does not apply.

124. The same conclusion follows also from a consideration of the principle of energy.

Since the energy of such a field is $\int \frac{K}{8\pi} F^2 d\tau$, if we assume that the system of electrostatic stresses is actually that given by Maxwell and that the energy due to it is *localized* in the elements of volume of the dielectric, we have, in an element of volume $d\tau$, energy W, where

$$W = \frac{KF^2}{8\pi} d\tau,$$

and therefore
$$dW = \frac{KFdF}{4\pi} d\tau.$$

But this energy must be equal to the work done by the stresses given above.

To find an expression for this, let α, β, γ be the sides of the volume $d\tau$, so that $\alpha\beta\gamma = d\tau$. Then the increments of α, β, γ, due to the electrostatic stresses, will be say $\alpha d\epsilon_1$, $\beta d\epsilon_2$, $\gamma d\epsilon_3$. And the work done by the stresses will be

$$(1) \qquad \frac{KF^2}{8\pi} \beta\gamma\alpha d\epsilon_1 = \frac{KF^2}{8\pi} d\tau\, d\epsilon_1,$$

$$(2) \qquad -\frac{KF^2}{8\pi} \gamma\alpha\beta d\epsilon_2 = -\frac{KF^2}{8\pi} d\tau\, d\epsilon_2,$$

$$(3) \qquad -\frac{KF^2}{8\pi} \alpha\beta\gamma d\epsilon_3 = -\frac{KF^2}{8\pi} d\tau\, d\epsilon_3;$$

$$\therefore\ dW = \frac{KFdF}{4\pi} d\tau = \frac{KF^2}{8\pi} d\tau\, (d\epsilon_1 - d\epsilon_2 - d\epsilon_3),$$

$$\therefore\ \epsilon_1 - \epsilon_2 - \epsilon_3 = \log F^2 + \text{const.}^*$$

This equation cannot be satisfied, if we regard electrostatic phenomena as essentially statical. For, if we suppose

$$\epsilon_1 = \epsilon_2 = \epsilon_3 = 0, \text{ when } F = F_0, \text{ we have } \epsilon_1 - \epsilon_2 - \epsilon_3 = \log \frac{F^2}{F_0^2},$$

and this is obviously admissible†.

125. It follows therefore that the distribution of stress given in Art. 121 is inadmissible,

(1) if the energy is assumed to be localized in elements of volume $d\tau$ of the field; or

(2) if the stresses are to be such as to satisfy the condition of the stability of the medium—the medium being assumed to be at rest and to obey the ordinary laws of elasticity.

* Poincaré, *Électricité et Optique*, Art. 83.

† With a further limitation on the arbitrariness of the quantities ϵ_1, ϵ_2, ϵ_3.

Recalling, further, the manner of obtaining the result in Art. 118, we conclude that, if the properties of the electric field are those of a strained elastic medium, the principal axes of stress are not those assumed in Art. 122; in other words, a line of force through a point is not a line of principal stress.

126. It follows, therefore, what indeed experiments seem to indicate, that the distribution of electrostatic stress in a dielectric medium is not so simple as Maxwell supposed, so that the assumption, virtually made by him, that the principal axes of stress (agreeably to Faraday's views) are along and perpendicular to the lines of force seems to be only a first approximation.

127. Experiments moreover suggest, as has been remarked by Sir J. J. Thomson, that there must be forces in the electric field not recognized by Maxwell's theory. These peculiarities of the field seem to be intimately related to the fact that the field is essentially one of *kinetic energy*. Some account has also to be taken of the 'charge' in the conductor.

128. If, however, we could assume that the energy of strain is kinetic energy. that the 'electrostatic' electromotive intensity produces kinetic phenomena (for instance, vortex motion), we could reconcile these contradictions.

129. The system of stresses obtained in Art. 118 or Art. 122 is however practically that which was qualitatively postulated by Faraday, who stated his views as follows:

"The direct inductive force, which may be conceived to be exerted in lines between two limiting and charged conducting surfaces, is accompanied by a lateral or transverse force, equivalent to a dilatation or repulsion of these representative lines."

130. As we know, the various electrical effects of attraction, repulsion, and induction can be easily explained in terms of these properties of the lines of force. Moreover, whatever be the nature of the constraint, it is obvious that the medium must be thrown into a certain polarised state* by the constraint, that is, a certain forced state due to electromotive force,

* Assuming this effect to be entirely statical.

the particles (of the medium), in the language of Faraday, assuming positive and negative points or parts, which are symmetrically arranged with respect to each other and the inductive surfaces or particles.

131. If, therefore, we admit that the electrical action is exerted by means of the medium intervening between electrified bodies, this medium must be in a state of constraint, although we are not justified in stating that the magnitudes and directions of these stresses are those obtained in Art. 122.

132. The expression for energy in terms of electric displacement is obviously given by

$$W = \frac{K}{8\pi}\int F^2 d\tau = \frac{2\pi}{K}\int \sigma^2 d\tau.$$

Moreover, the quantity σ, being proportional to F, is a vector. Let its resolutes be f, g, h; then we have

$$K\frac{\partial V}{\partial x} = -4\pi f, \text{ etc., and } W = \frac{2\pi}{K}\int (f^2 + g^2 + h^2)\, d\tau.$$

Also, since

$$\frac{\partial}{\partial x}\left(K\frac{\partial V}{\partial x}\right) + \ldots + \ldots = 0 \text{ or } -4\pi\rho,$$

we have $$\frac{\partial f}{\partial x} + \frac{\partial g}{\partial y} + \frac{\partial h}{\partial z} = 0 \text{ or } \rho,$$

and if $\dot{f} \equiv \frac{df}{dt}$, etc.,

$$\frac{\partial \dot{f}}{\partial x} + \frac{\partial \dot{g}}{\partial y} + \frac{\partial \dot{h}}{\partial z} = 0 \text{ (in a stationary medium, if } \rho \text{ or } \dot{\rho} = 0),$$

where ρ is the volume density of *free* electricity at any point. In other words, electric displacement in space unoccupied by electricity is subject to the condition that the medium is incompressible as regards such displacements*, so that the movement of electricity giving rise to electric displacement and to electric current [Art. 139] in such a medium obeys the law of an incompressible fluid*.

* Observing that in an elastically incompressible medium $\frac{\partial u}{\partial x} + \ldots + \ldots = 0$, where u, v, w are displacements, and in an incompressible fluid, the same equation holds, u, v, w being now velocities.

133. The distinction between 'free electricity' and electric displacement is important; we have, in fact, the general expression for a dielectric medium

$$\frac{\partial f}{\partial x}+\frac{\partial g}{\partial y}+\frac{\partial h}{\partial z}=\rho,$$

where ρ is the volume density of free electricity as in Art. 121 and the electric property of such a medium will be determined by the above equation, while its dielectric constant will depend on the quantity ρ. It will be observed that the electron theory is formally based on this hypothesis of Maxwell. [Art. 233.]

134. Further, if we describe a surface S round a charge e, then

$$4\pi e=\int KF_n dS=4\pi\int\sigma dS,$$

or

$$e=\int\sigma dS,$$

that is, an electric charge is equal to the surface integral of normal electric displacement over any closed surface surrounding the charge.

135. It follows therefore that the total quantity of electric displacement within a closed surface always remains the same, so that the motion of electricity is like that of an incompressible fluid [Art. 132].

136. Maxwell thus illustrates the physical meaning of electric displacement:

Let us consider an accumulator formed of two conducting plates A and B separated by a dielectric. Let the two poles of a voltaic cell be joined to A and B. Then A and B will receive certain charges. This is effected by the passage of a current by means of what may be described as a real transference of 'electricity*.' A, B in this way come to be at different potentials. But at the same time that this takes place the electromotive force between A and B acting on the dielectric between A and B produces an electric displacement of the same magnitude as the quantity of electricity that is transferred during the passage of the current. The motion of the electricity thus takes place in a closed circuit. [See, however, Art. 143.]

* Actual transference of positive particles and electrons in opposite directions. [Art. 256.]

137. We may thus conveniently distinguish two kinds of currents, a conduction current and a displacement or polarisation current. And since the strength of an electric current is measured by the quantity of electricity that passes across a cross-section in unit time, the displacement or polarisation current will, on the same principle, be represented by $\dot{\sigma}\,(=\dot{f}, \dot{g}, \dot{h})$, i.e. the time rate of change of σ, if σ is the amount of electric displacement. We may distinguish other kinds of currents also. These we proceed to consider.

138. Ordinarily, a gas is a non-conductor, but it may be put into the conducting state by various means, such as exposure to X-rays. This is due to the presence in the gas of electrified particles or ions which are produced by these means, and the gas is said to be ionised. Moreover, when rapidly moving ions [Ch. IV] pass through a gas and come into collision with its molecules, the gas is further ionised and a current passes. Conduction through a gas, therefore, is in reality a procession of ions or charged particles through the gas. Such a current may be called a convection current*.

139. An electric current may thus be expressed provisionally by the following equations:

$$\left.\begin{aligned}\dot{\sigma} &= \text{displacement current}\\ K &= \text{specific inductive capacity}\end{aligned}\right\}\ F_1 = \frac{K}{4\pi}\,\dot{\sigma},$$

$$\left.\begin{aligned}i &= \text{Ohmic current}\\ R &= \text{resistance of the circuit}\end{aligned}\right\}\ F_2 = Ri \text{ or } i = \frac{F_2}{R},$$

$$I = \text{ionisation current},$$

where $I = e\,(q_1 n_1 + q_2 n_2)\,F_3$,

e = charge on an ion,

q_1, q_2 = velocities of +, − ions, under unit electric intensity,

n_1, n_2 = number of ions per unit volume,

F_1, F_2, F_3 being the corresponding electromotive intensities.

140. One point of difference between the displacement and conduction currents on the one hand and the convection

* J. J. Thomson, *Conduction of Electricity through Gases.*

current on the other is that the E.M.F. is not necessarily proportional to the current in the case of the latter, whereas it is so in the former.

It will be seen, however, on examination that there is no real distinction between them.

141. To fix our ideas:—Consider an electrical system consisting of a voltaic cell, the poles of which are connected to two conducting wires ending in two parallel plates with an air-gap. When the difference of potentials is greater than a certain amount*, ions (+, −) are thrown off in sufficient numbers to produce a steady current across the air-gap, and when the difference of potentials is great enough to produce sufficiently copious ionisation, the negative ions or corpuscles moving with sufficient velocity produce luminosity of the gas and a discharge passes.

142. Thus, the path of an electric discharge, silent or luminous, is the path of a current of electricity in the same sense in which we use the expression, "a current of electricity" in a closed conducting circuit joining the poles of a battery, and, therefore, since an electric discharge is obviously due to a movement of ions, an electric current must be a "procession and not an arrangement." In the voltaic cell itself, there is an actual procession going on of ordinary ponderable atoms which must be carriers of electricity. In the same way, a Leyden jar represents a continuous circuit for the flow of electricity.

143. We may thus modify, in a general manner, Maxwell's theory of electric displacement in view of the theory of ionisation in gases.

As soon as the plates in Art. 136 are connected with the poles of the battery, by the action of the E.M.F. in the latter, a transference of electricity takes place in the wire charging one of the plates, say, positively, and the other negatively, and producing a difference of potential between the plates.

Across the dielectric, on the other hand, an electric

* Even before this—even for the slightest potential difference, it is likely that in the intervening air, there is a current or procession of ions, due to spontaneous ionisation.

displacement *and* ionisation take place, the total amount of which is equivalent to the electricity transferred along the wire. The transfer or procession therefore takes place in a closed circuit.

When the plates are discharged by disconnecting them from the battery and connecting them up by a wire, the displacement and ionisation will be in the opposite direction.

In either case, there is a current due to a procession of electricity, but this current is necessarily transient. When an E.M.F. continues to act after the current has once passed, a state is reached on account of the electric elasticity of the medium which is one of dynamic stability. The electric displacement thus seems to be associated with what is not of the nature of a static but of a kinetic phenomenon.

144. The state, in fact, is one of convective equilibrium. There is generation and re-combination of positive ions and corpuscles, so that there is always a current. This is small, unless the E.M.F. of the battery is sufficiently great, when ionisation may be sufficient for a strong and steady current, with or without luminosity.

145. On the whole, therefore, every case of electric current is that of procession of ions—in electric or metallic conduction as well as in dielectric polarisation or gaseous convection; while every case of electrification must be held to consist in a passage of a current, instantaneous, transient or continuous.

SECTION II. MAGNETISM

146. The phenomena of magnetism can also be explained by postulating the existence of stresses in the *magnetic* field, as in the case of electric phenomena. This we proceed to consider.

147. Since any permanent magnet can be split up into elementary magnets, whose axes are in the direction of magnetisation, the potential energy of a magnet in a field, whose potential at any point is V, is $\Sigma M \frac{\partial V}{\partial s}$, where M is the magnetic moment of an elementary magnet at any point, and ds = the length of the elementary magnet.

And since $M = Id\tau$, I being the intensity of magnetisation and $d\tau$ the volume of the elementary magnet,

$$\text{The potential energy} = W = \int I d\tau \frac{\partial V}{\partial s}$$

$$= \int I \left(\lambda \frac{\partial V}{\partial x} + \mu \frac{\partial V}{\partial y} + \nu \frac{\partial V}{\partial z}\right) d\tau,$$

if λ, μ, ν define the direction of magnetisation,

$$= \int \left(A \frac{\partial V}{\partial x} + B \frac{\partial V}{\partial y} + C \frac{\partial V}{\partial z}\right) d\tau$$

$$= \int (Al + Bm + Cn)\, dS - \int V \left(\frac{\partial A}{\partial x} + \dots + \dots\right) d\tau$$

$$= \int I \cos\theta\, V\, dS - \int V \left(\frac{\partial A}{\partial x} + \dots + \dots\right) d\tau \quad \dots\dots\dots(1),$$

where dS is an element of surface of the magnet, θ = the inclination of the normal (l, m, n) at dS to the direction of magnetisation, while A, B, C are components of magnetisation at any point (x, y, z).

Now, if we take

$$I \cos\theta = \sigma, \quad \text{and} \quad -\left(\frac{\partial A}{\partial x} + \frac{\partial B}{\partial y} + \frac{\partial C}{\partial z}\right) = \rho,$$

we get $$W = \int \sigma V dS + \int \rho V d\tau \dots\dots\dots\dots\dots\dots\dots\dots(2)$$

from (1).

Or, the energy may be conceived to be due to a surface distribution $= I \cos\theta$, and a volume distribution $= -\left(\frac{\partial A}{\partial x} + \dots + \dots\right)$ of magnetism.

148. Suppose now that the potential V is due to the magnet itself to which σ, ρ refer. Then, since the law of force is the same as in the case of electric phenomena, V will satisfy the conditions

$$\nabla^2 V + 4\pi\rho = 0 \dots\dots\dots\dots\dots\dots\dots\dots(3),$$

$$-\left(\frac{\partial V}{\partial n}\right)_1 dS_1 + 4\pi\sigma dS_1 + \left(\frac{\partial V}{\partial n}\right)_2 dS_2 = 0 \quad \dots\dots\dots(4),$$

where dS_2 is an element of the surface of the magnet just outside σ.

From (3) we have

$$\int \nabla^2 V d\tau - 4\pi \int \rho d\tau = 0, \quad \text{or} \int \left(\frac{\partial a}{\partial x} + \ldots + \ldots\right) d\tau = 0 \ldots(5),$$

if we write $\quad a = -\dfrac{\partial V}{\partial x} + 4\pi A = \alpha + 4\pi A$, etc.

Again, (4) may be rewritten

$$\int \frac{\partial V}{\partial n} dS - 4\pi \int (Al + Bm + Cn)\, dS$$

$$= \int \left(\frac{\partial V}{\partial n} - 4\pi I \cos\theta\right) dS = 0 \ldots(6).$$

If we now put

$$B_n = H_n + 4\pi I \cos\theta = -\frac{\partial V}{\partial n} + 4\pi I \cos\theta \ldots\ldots(7),$$

i.e. if H_n = the magnetic force in the direction of the normal, we have (over any closed surface) the condition

$$\int B_n dS = 0 \quad \ldots\ldots\ldots\ldots\ldots\ldots\ldots(8),$$

or the total magnetic induction over any closed surface is zero, if we define $B_n dS$ to be the normal magnetic induction over dS, B_n being the normal component of a, b, c, i.e. $= al + bm + cn$.

Again, (4) gives (attending to the signs with obvious notations)

$$dS_1 ({}_1H_n + 4\pi I \cos\theta) = {}_2H_n dS_2 \quad \ldots\ldots\ldots\ldots(9),$$

or

$$dS_1 \,.\, {}_1B_n = dS_2 \,.\, {}_2B_n \quad \ldots\ldots\ldots(10),$$

if the medium (2) separated by dS_1 from the medium (1) has no magnetism, permanent or induced, in the neighbourhood of dS.

149. Now (7) is equivalent to the statement that the resultant of H and I is B, which is therefore the total resultant magnetic induction.

Draw a small tube inside the magnet having its generating lines in the direction of resultant magnetic induction. Let us call this a tube of magnetic induction. Then, applying (8) to this tube, we get

$$B_n dS + B_n' dS' = 0.$$

Writing this in the form

$$B_n dS = -B_n' dS',$$

we conclude that the normal induction along such a tube of magnetic induction drawn inside a surface is constant and is always directed *in the same way*, remembering the meaning of $\frac{\partial V}{\partial n}$ in (6).

Then (10) shows that the same is true for a tube drawn outside.

150. We conclude, therefore, that (*a*) a tube of induction forms a close circuit both inside and outside a magnetic body; (*b*) the normal induction is constant throughout such a tube; (*c*) the tube of induction coincides with the tube of magnetic force outside a magnet, that is, in space, devoid of magnetism.

Note that the lines of magnetic force are in opposite senses inside and outside the magnet.

151. If the medium surrounding the magnet is filled with a magnetic substance (all substances are more or less so), there will be induced magnetism on it, and the equation (10) will still be true; only the magnetisation in the second medium*, ${}_2I_n$ (say), being induced, it must be in the direction of H_n: that is

$${}_2I_n = {}_2H_n \,.\, \kappa_2,$$

where κ_2 is constant (provided the magnetising force is small), the nature of which depends on the nature of the medium;

$$\therefore\quad {}_2B_n = {}_2H_n(1 + 4\pi\kappa_2) = \mu \,.\, {}_2H_n \text{ (say)},$$

where μ is a constant, if ${}_2H_n$ is small, but which in any case depends on the nature of the medium as well as on the magnitude of H and also in some cases on its direction (as in magne-crystallic phenomena). μ is called magnetic permeability.

Along a tube of induction, therefore, we have

$B_n dS =$ constant and for a magnetic medium, $\mu H_n dS =$ constant.

At the surface of separation of two magnetic media (μ, μ')

$$\mu H_n = \mu' H_n'.$$

If one of the media is a permanent magnet, μH_n for that medium stands for

$$H_n + 4\pi I_n.$$

152. The properties of a tube of induction are therefore mathematically equivalent to the properties of a tube of electric induction and we conclude that the magnetic field may be mapped out metrically in the same way as an electric field.

* Assuming it to be uniformly magnetic.

Again, the energy of a magnetic system due to *the mutual* action of its parts is $\frac{1}{2}\Sigma mV$, where m is the strength of a permanent magnetic pole.

Hence, this energy $= \frac{1}{2}\Sigma m \frac{\partial V}{\partial s} ds = \frac{1}{2}\Sigma M \frac{\partial V}{\partial s}$, where ds is the length of an elementary magnet and M = its magnetic moment.

Therefore, if I_1 = strength of permanent magnetisation, the energy $E = \frac{1}{2}\int I_1\, d\tau\,(-H)$, $d\tau$ being an elementary volume of a permanent magnet;

$$\therefore\quad E = -\tfrac{1}{2}\int (A_1\alpha + B_1\beta + C_1\gamma)\, d\tau,$$

where A_1, B_1, C_1 are the resolved parts of I_1 and α, β, γ those of H.

But if a, b, c are the resolved parts of B, from Art. 148

$$a = \alpha + 4\pi A = \alpha + 4\pi(A_1 + \kappa\alpha) = \mu\alpha + 4\pi A_1;$$

$$\begin{aligned}\therefore\quad E &= -\frac{1}{8\pi}\int \{(a - \mu\alpha)\,\alpha + \ldots + \ldots\}\, d\tau \\ &= \frac{1}{8\pi}\int \mu H^2 d\tau - \frac{1}{8\pi}\int (a\alpha + b\beta + c\gamma)\, d\tau \\ &= \frac{1}{8\pi}\int \mu H^2 d\tau,\end{aligned}$$

since $\displaystyle\int (a\alpha + b\beta + c\gamma)\, d\tau = -\int \left(a\frac{\partial V}{\partial x} + \ldots + \ldots\right) dx\,dy\,dz$

$$= -\int V B_n dS + \int V \left(\frac{\partial a}{\partial x} + \ldots + \ldots\right) dx\,dy\,dz = 0,$$

for the surface integral reduces to terms at each element of a surface of separation and therefore necessarily vanishes, while at each point, we have the *solenoidal* condition

$$\frac{\partial a}{\partial x} + \ldots + \ldots = 0 \text{ [Art. 148]}.$$

Therefore, exactly as in explaining electrostatic energy, we may postulate a distribution of magnetic tubes of stress in explaining magnetic energy.

SECTION III. ELECTRO-MAGNETISM

153. The experiments of Œrsted prove that an electric current produces a magnetic field. If, then, we admit that an electric current consists in a procession of charged corpuscles, since such a procession of corpuscles must carry lines of force with them, we conclude that the motion of lines of force gives rise to or is associated with a magnetic field in which magnetic lines of induction are distributed according to the law already found.

Further, since all cases of charge and discharge of electricity and, therefore, all electrical phenomena consist, on Maxwell's view, in motion of electricity in closed circuits, we may describe an electric field as follows:

154. In every part of an electric field, there is a procession of electric charges; in a dielectric, it is transient (being, however, still associated with ethereal motion, if we accept the view that all energy is kinetic); while the procession lasts, it gives rise to magnetic induction in every part of the field. The whole field, moreover, is filled with lines of electric induction, which move with the motion of electricity. The field is under stress, imaged by electric and magnetic lines of induction.

155. Now, the magnetic potential at any point due to a magnetic shell is $\phi\Omega$, where ϕ is the strength of a shell and Ω the solid angle, subtended by the shell, at the point.

Experiments of Ampere and Weber show that a small circuit carrying a current behaves as a magnetic shell. The potential due to it will therefore be of the same form except for a constant factor. Since a circuit of finite dimensions may be conceived to be made up of elementary circuits, the potential due to a finite circuit must also be of the same form. But as the space surrounding a wire is a cyclic space, while the space surrounding a magnetic shell is a-cyclic, the potential due to a current must necessarily be a cyclic function, while the other potential is single-valued. This is verified by experiment.

156. Again, experiment shows that the force due to an infinite current i along the axis of z, at a distance r from the current, is $\dfrac{2i}{r}$.

The corresponding potential will therefore be

$$2i \tan^{-1}\frac{y}{x} + \text{constant},$$

a cyclic function.

Identifying this with a magnetic shell, coinciding with the infinite plane, of which the infinite current is the boundary, we conclude that the potential due to a circuit carrying a current i

$$= i\Omega + 4s\pi i$$

at any point, where Ω is the solid angle, subtended by the circuit at the point, and s, an integer.

157. It follows, moreover, that the mechanical force in the direction of x, between a circuit carrying a current of strength i and a magnetic system, is $i\dfrac{\partial N}{\partial x}$, where $N = \int B_n dS$*, B_n being due to the external magnetic system and dS an element of the shell with which the circuit is identified.

Since the total magnetic induction over any closed surface is zero, induction across any surface depends only on the bounding line. Hence the surface integral can be replaced by a line integral.

We may, therefore, write

$$\int B_n dS = \int (F dx + G dy + H dz),$$

where

$$B_n = la + mb + nc.$$

Hence it follows easily that $a = \left(\dfrac{\partial H}{\partial y} - \dfrac{\partial G}{\partial z}\right)$, etc., for obviously

$$\int \left[\left(\frac{\partial H}{\partial y} - \frac{\partial G}{\partial z}\right) l + \ldots + \ldots\right] dS = \int \left[\left(\frac{\partial H}{\partial y} - \frac{\partial G}{\partial z}\right) dy\,dz + \ldots + \ldots\right]$$

$$= \left\{\int \left[- H dz + \frac{\partial}{\partial y}(H dz)\, dy + H dz - \frac{\partial}{\partial x}(H dz)\, dx\right] + \ldots + \ldots\right\}$$

$$= \int F dx + G dy + H dz.\dagger$$

If, now, W is a function, such that the attraction P in the direction of x is $\dfrac{\partial W}{\partial x}$ (between two circuits of invariable form), then

$$P = i\frac{\partial N}{\partial x} = ii'\frac{\partial M}{\partial x},$$

where

$$Mii' = W = i\int (F' dx + G' dy + H' dz),$$

F', G', H' referring to i'.

* Since the shell does not belong to the magnetic system, magnetic force and magnetic induction coincide. Moreover, the magnetic induction like the electric induction is given by the number of tubes of induction [Maxwell, *Elect. and Mag.* Art. 489].

† *Ibid.* Art. 24, Vol. I.

If u, v, w be the densities of a current i in three rectangular directions, defined by the relation

$$idx = udxdydz, \text{ etc.,}$$

then $$W = \int (F'u + G'v + H'w)\, dxdydz.$$

158. We have seen that a circuit carrying a current creates a magnetic field in its neighbourhood.

If V = the magnetic potential due to the current and Ω = the solid angle subtended by the circuit at any point, then

$$V = i\Omega + 4\pi si \quad \text{.....................(1),}$$

i.e. $$\int (\alpha dx + \beta dy + \gamma dz) = 4\pi i = 4\pi \int (lu + mv + nw)\, dS \ldots (2),$$

where α, β, γ are the resolved magnetic forces in the directions of x, y, z, the line integral being taken round a line of magnetic force, and the surface integral over a surface, of which the line is the periphery.

Then, obviously we have as in Art. 157

$$4\pi u = \frac{\partial \gamma}{\partial y} - \frac{\partial \beta}{\partial z} \quad \text{.....................(3),}$$

which shows further that

$$\frac{\partial u}{\partial x} + \ldots + \ldots = 0,$$

so that the current (u, v, w) may be regarded as that of an incompressible fluid.

159. Referring to Art. 139, we observe that the current in general will consist of three parts:—(1) polarisation current $(\dot{f}, \dot{g}, \dot{h})$, (2) ohmic current (say) (p, q, r), (3) ionisation current. Ignoring the separate existence of (3), for the present, in accordance with Maxwell's theory, we get

$$u = p + \dot{f}.$$

160. Again, since the energy of a magnetic field is

$$\frac{1}{8\pi} \int \mu H^2 d\tau,$$

this must be the energy of such a field, if the field is due to currents, and, as such, should [Art. 142] be called electro-kinetic energy.

This is

$$=\frac{1}{8\pi}\int(a\alpha+b\beta+c\gamma)\,d\tau$$

$$=\frac{1}{8\pi}\int\left\{\alpha\left(\frac{\partial H}{\partial y}-\frac{\partial G}{\partial z}\right)+\ldots+\ldots\right\}d\tau,\quad\text{[Art. 157]},$$

$$=\frac{1}{8\pi}\int\left\{F\left(\frac{\partial\gamma}{\partial y}-\frac{\partial\beta}{\partial z}\right)+\ldots+\ldots\right\}d\tau$$

$$=\frac{1}{2}\int(Fu+Gv+Hw)\,d\tau,\ \text{from (3) of Art. 158},$$

F, G, H being due to the total field [Art. 157].

The vector (F, G, H) is evidently of the nature of a momentum and may be called the electro-kinetic momentum, if we can regard (u, v, w) as a velocity. But this, as we have seen, we are justified in doing. This vector is called the vector-potential of magnetic induction.

161. Now
$$\mu\alpha=\frac{\partial H}{\partial y}-\frac{\partial G}{\partial z},\ \text{etc.},$$

and
$$4\pi u=\frac{\partial\gamma}{\partial y}-\frac{\partial\beta}{\partial z},\ \text{etc.};$$

$$\therefore\quad 4\pi\mu u=\frac{\partial}{\partial x}\left(\frac{\partial F}{\partial x}+\frac{\partial G}{\partial y}+\frac{\partial H}{\partial z}\right)-\nabla^2F.$$

Let $F=F_1+F_2$, such that

$$\nabla^2F_1+4\pi\mu u=0,\ \text{etc.},$$

and
$$\frac{\partial F_1}{\partial x}+\frac{\partial G_1}{\partial y}+\frac{\partial H_1}{\partial z}=0.$$

Then
$$-\nabla^2F_2+\frac{\partial}{\partial x}\left(\frac{\partial F_2}{\partial x}+\frac{\partial G_2}{\partial y}+\frac{\partial H_2}{\partial z}\right)=0,$$

which shows that
$$F_2=\frac{\partial\chi}{\partial x},\ \text{etc.},$$

where χ is an arbitrary function; therefore, all the equations will be satisfied if we take

$$F_1=\int\frac{\mu u'}{r}\,dx'dy'dz',\ \text{and}\ F_2=\frac{\partial\chi}{\partial x},\ \text{etc.}$$

But F_2 does not enter into the expressions for a, b, c.

Moreover,

$$\frac{\partial F_1}{\partial x} + \ldots + \ldots = \mu \frac{\partial}{\partial x}\int \frac{u'}{r}\, dx'dy'dz' + \ldots + \ldots$$

$$= \mu\int\left(\frac{\partial u'}{\partial x} + \ldots + \ldots\right)\frac{1}{r}\, dx'dy'dz' - \mu\int\left(u'\frac{\partial}{\partial x'}\frac{1}{r} + \ldots + \ldots\right)dx'dy'dz' = 0,$$

provided μ is independent of the current (which, however, is not admissible), since $(x-x')^2 + (y-y')^2 + (z-z')^2 = r^2$,

$$\frac{\partial}{\partial x}\frac{1}{r} = -\frac{\partial}{\partial x'}\frac{1}{r},\quad l'u' + m'v' + n'w' = 0, \text{ and } \frac{\partial u'}{\partial x} + \ldots + \ldots = 0.$$

On the understanding, however, that μ *is* independent of the current, F_1 may be taken equal to F. Accordingly,

$$F_1 = F = \mu\int\frac{u'dx'dy'dz'}{r} = \mu\int\frac{u}{r}\, dxdydz = \mu\int\frac{idx}{r}\,*.$$

162. Now, the energy of an electric circuit is due not merely to the field created by other electric systems, but also to that created by itself.

Hence, the energy of the circuit (supposed to be at rest) due to its own current is of the form Ni, where

$$Ni = i\int B_n dS = i\int(la + mb + nc)\, dS$$

$$= i\int(Fdx + Gdy + Hdz)$$

$$= \frac{i^2}{2}\int\left(\frac{dxdx' + \ldots + \ldots}{r}\right),$$

remembering that each element dx or dx' belonging to the same circuit occurs twice.

The total electro-kinetic energy† due to currents in more than one circuit is therefore T, where

$$2T = L_1 i_1^2 + 2M_{12} i_1 i_2 + \ldots,$$

and
$$M_{12} = \int\frac{dx_1 dx_2 + \ldots + \ldots}{r_{12}} = \int\frac{\cos\omega_{12}}{r_{12}}\, ds_1 ds_2, \text{ etc.}$$

ω_{12} being the angle between the elements ds_1, ds_2 of the circuits (1, 2) and r_{12} the distance between them.

* Observe that $\frac{\partial F}{\partial x} + \ldots + \ldots = 0$ and $\frac{\partial u}{\partial x} + \ldots + \ldots = 0$ correspond to the assumption $\Delta = 0$ of the elastic solid theory.

† Since this energy depends on the configuration of the system, any motion tending to change the configuration will also tend to change the energy. Regarded from this point of view, the energy may be called electro-dynamic energy.

163. Consider now the flow of energy in any circuit of the system; since the energy due to a current i must be equal to the sum of

(1) the energy lost in frictional generation of heat, and

(2) increase of electro-dynamic energy [Art. 162, note],

we have $$Eidt = \sigma i^2 dt + idN,$$

where E is the impressed E.M.F. in the circuit, σ the resistance.

If there is no impressed E.M.F. in the circuit,

$$\sigma i = -\frac{dN}{dt} = \int (X dx + Y dy + Z dz),$$

if the electromotive intensity $= (X, Y, Z)$.

$$\therefore \quad -\frac{d}{dt}\int (la + mb + nc)\, dS = \int (X\, dx + Y\, dy + Z\, dz)$$

$$= -\frac{d}{dt}\int (F dx + G dy + H dz),$$

$$\therefore \quad X = -\frac{\partial F}{\partial t} - \frac{\partial \psi}{\partial x},$$

where ψ is an arbitrary function which is introduced to make the solution general. To interpret this, we notice that

$$\nabla^2 \psi + 4\pi\rho = 0,$$

i.e. ψ is the potential of free electricity of density ρ.

We have also

$$\frac{\partial Z}{\partial y} - \frac{\partial Y}{\partial z} = -\dot{a}, \text{ etc. [Art. 157] } \quad \text{..............(1)},$$

the medium being stationary.

Again, from Art. 159 we get

$$u = \frac{X}{\sigma} + \frac{4\pi}{K}\dot{X},$$

and from Art. 158 $$4\pi\mu u = \frac{\partial c}{\partial y} - \frac{\partial b}{\partial z} \quad \text{..................(2)}.$$

Thus, for a conductor $$u = \frac{X}{\sigma},$$

and for a dielectric $$u = \frac{K}{4\pi}\dot{X} \quad \text{......................(3)}.$$

164. From (1), (2), (3) of Art. 163, if μ is constant, we get $\mu K\ddot{X} = \nabla^2 X$; that is, in a medium ($\mu$, K) the electric force is propagated with the velocity $\frac{1}{\sqrt{\mu K}}$.

In the electro-magnetic system $\mu = 1$, $K = \frac{1}{v^2}$ for air; where $v =$ the ratio of the electrostatic to the electro-magnetic units.

Therefore, in air, the velocity of propagation of an electro-magnetic disturbance should be equal to v, while from direct experiment we find that the velocity of light is equal to this ratio.

We are, therefore, justified in concluding that light is an electro-magnetic disturbance.

165. If n is equal to the index of refraction, we have $n^2 = \mu K$; and as μ is very nearly the same in all non-magnetic media, we deduce that $n^2 = K$.

In testing the accuracy of the relation, we have to remember that n refers to infinitely short electro-magnetic vibrations which constitute light and therefore K should be determined in relation to such infinitely short waves. A comparison of calculated results with experimental data, therefore, cannot be instituted in the form in which theory requires it. In spite of this, n^2 for the longest light waves experimented upon has been found to agree with K for slowly varying fields in several instances, although it differs in others*, while a direct determination of the velocity of propagation of electro-magnetic disturbance and that of K have also yielded results which are found to agree in many cases.

166. Moreover, as n depends on wave-length, while K appears as a constant (independent of wave-length) in the

	n for yellow light	$\sqrt{K}$
* Air	1·000294	1·000295
Hydrogen	1·000138	1·000132
CO_2	1·000449	1·000473
NO	1·000346	1·000345
H_2O	1·333 at 0°	·9
Methyl alcohol	1·3379 at 0°	5·7

above investigation, it is obvious that the theory will require modification, in order that there may be agreement between the results of theory and experiment.

Direct experiments in verification of the theory that light is an electro-magnetic disturbance were supplied by the researches of Hertz, who showed that an oscillatory electric discharge (the nature of which had been previously discussed by Lord Kelvin) induces varying electromotive forces in the field, which can be detected by means of a suitable circuit with a spark gap; in other words, an electro-magnetic oscillation is propagated with the velocity of light.

167. In order to illustrate this, consider the case of the discharge of a Leyden jar*.

Let Q = charge of the jar,
i = the current,
C = capacity,
R = the resistance of the circuit in connection with it,
L = the coefficient of self-induction.

Then, $$E - iR = \frac{d}{dt}(Li) \text{ [Art. 163]},$$

while $$i = -\frac{dQ}{dt}, \text{ from definition.}$$

$$\therefore \quad L\ddot{Q} + R\dot{Q} + \frac{Q}{C} = 0; \text{ therefore, if } \frac{1}{CL} > \frac{R^2}{4L^2},$$

the discharge is oscillatory and

$$T = \text{periodic time} = \frac{2\pi}{\sqrt{\frac{1}{CL} - \frac{R^2}{4L^2}}}.$$

If $\frac{R}{L}$ be negligible, then

$$T = 2\pi\sqrt{CL}.$$

Such a system may serve as a source of electro-magnetic disturbance.

168. Generally†, a wire with a spark gap will serve the same purpose, if we arrange a variable electromotive force to work in the system. In that case, it may be called a vibrator.

* W. Thomson, *Phil. Mag.* 4th series, v.

† Hertz, *Ausbreitung der elektrischen Kraft*; J. J. Thomson, *Recent Researches.*

A wire with a spark gap, if placed in a field of varying electromotive force, will exhibit oscillatory discharge if its natural period of oscillation coincides with the period of oscillation of the source. Such an arrangement may therefore be called a resonator.

If the source has a metallic (zinc) reflector behind it, the maximum effect is manifested, for the superposition of direct and reflected waves causes stationary waves, and at definite points in the region, intervening between the source and the detector, there are, as in the case of sound, loops and nodes. The electro-magnetic vibrations produce other effects also. Falling on loose iron filings, they make them *cohere*; that is, if these filings form part of the circuit of an electric current, they allow the passage of the current through them, every time the vibrations impinge on them. An electric circuit completed through a tube containing loose iron filings and enclosing a galvanometer may also, therefore, effectively serve as a detector.

With a suitable source and detector, it has been found possible to verify that optical and electro-magnetic media are identical.

169. In order that T should be infinitesimal, we must have C and L infinitely small quantities of the first order and R an infinitesimal of the second order, so that the electro-magnetic system whose oscillatory discharge will set up vibrations of light must be of infinitesimal dimensions.

170. From the equations of Art. 163, we can derive generally

$$\mu K\ddot{\phi} = \nabla^2\phi \text{ (for a dielectric),}$$

where ϕ stands for any of the quantities

$$X, Y, Z, \quad a, b, c, \quad f, g, h, \quad u, v, w.$$

In particular, let us consider $\ddot{f} = V^2\nabla^2 f$, where $K\mu V^2 = 1$.

If we have plane waves defined by

$$f = f_0 \sin\frac{2\pi}{\lambda}(lx + my + nz - Vt), \text{ etc.,}$$

then, since

$$\frac{\partial f}{\partial x} + \dots + \dots = 0,$$

$$lf_0 + mg_0 + nh_0 = 0,$$

i.e. the electric displacement is in the wave front.

And also since

$$-\mu\dot{\alpha} = \frac{\partial Z}{\partial y} - \frac{\partial Y}{\partial z} = \frac{4\pi}{K}\left(\frac{\partial \dot{h}}{\partial y} - \frac{\partial \dot{g}}{\partial z}\right)$$

$$= \frac{4\pi}{K} \cdot \frac{2\pi}{\lambda}(mh_0 - ng_0) \sin \frac{2\pi}{\lambda}(lx + my + nz - Vt),$$

we have $$\alpha = 4\pi V(ng - mh), \text{ etc.,}$$

or $$l\alpha + m\beta + n\gamma = 0.$$

That is, the magnetic force (α, β, γ) is also in the wave front.

We have, therefore, in the wave front, both electric displacement and magnetic force, perpendicular to each other.

There is, thus, in the wave front, two sets of effects at right angles to each other, one set in the plane of polarisation and the other perpendicular to it. The question as to which of them may be taken to correspond to the elastic displacement of the elastic solid theory cannot, therefore, be uniquely decided.

171. Let us now consider the flow of energy in the electromagnetic field. The electrostatic energy W in a dielectric is

$$= \tfrac{1}{2}\int (Xf + Yg + Zh)\, d\tau,$$

where $d\tau$ = element of volume.

$$\therefore \quad \delta W = \tfrac{1}{2}\int (X\delta f + Y\delta g + Z\delta h)\, d\tau$$
$$+ \tfrac{1}{2}\int (f\delta X + \dots + \dots)\, d\tau = \int (X\delta f + \dots + \dots)\, d\tau,$$

since $$KX = 4\pi f, \text{ etc.}$$

$$\therefore \quad \dot{W} = \int (X\dot{f} + Y\dot{g} + Z\dot{h})\, d\tau.$$

The heat evolved by Ohm's law is H, where

$$\dot{H} = \int \sigma (p^2 + q^2 + r^2)\, d\tau,$$

and (p, q, r) is the conduction current, so that

$$\dot{H} = \int (Xp + Yq + Zr)\, d\tau, \text{ since } \sigma p = X.$$

Finally, the electro-kinetic energy is T, where

$$T = \frac{\mu}{8\pi}\int (\alpha^2 + \beta^2 + \gamma^2)\, d\tau,$$

so that

$$\dot{T} = \frac{\mu}{4\pi}\int (\alpha\dot{\alpha} + \beta\dot{\beta} + \gamma\dot{\gamma})\, d\tau = \frac{1}{4\pi}\int (a\dot{a} + \ldots + \ldots)\, d\tau,$$

provided we can take $\mu\dot{\alpha} = \dot{a}$.

Therefore, the rate of increase of energy in a stationary element of volume is $\dot{W} + \dot{H} + \dot{T}$.

This ought to be expressible as a surface integral, if the energy resides in the medium.

Now (the same system of units being used for all the quantities)

$$\dot{W} + \dot{H} + \dot{T} = \frac{1}{4\pi}\int \left\{\alpha\dot{a} + X\left(\frac{\partial\gamma}{\partial y} - \frac{\partial\beta}{\partial z}\right) + \ldots + \ldots\right\} d\tau,$$

since $$4\pi (f + p) = \left(\frac{\partial\gamma}{\partial y} - \frac{\partial\beta}{\partial z}\right).$$

$$\therefore\ \dot{W} + \dot{H} + \dot{T} = \frac{1}{4\pi}\int dS\, \{(m\gamma - n\beta) X + \ldots + \ldots\}$$
$$+ \frac{1}{4\pi}\int d\tau \left[\left\{\alpha\dot{a} - \alpha\left(\frac{\partial Y}{\partial z} - \frac{\partial Z}{\partial y}\right)\right\} + \ldots + \ldots\right]$$
$$= \frac{1}{4\pi}\int dS\, \{(m\gamma - n\beta) X + \ldots + \ldots\},$$

since $\dfrac{\partial Y}{\partial z} - \dfrac{\partial Z}{\partial y} = \dot{a}$, etc. [Poynting, *Phil. Trans.* 1884.]

172. We thus find that the rate of increase of energy is equal to the normal component of a certain flux. The direction of this flow is evidently perpendicular to both the electric and the magnetic force, so that the direction of propagation of energy is perpendicular to these directions. This direction is, evidently, the direction of propagation of light, if optical energy is the same as the energy of electro-magnetic disturbance.

173. It has been assumed in this investigation that the energy is localized in each element of volume. Now both electrostatic and electro-magnetic energies have been shown to be capable of being estimated on this postulate and so also, the ohmic energy of electric current. We conclude, therefore, that the above theory is consistent with the other parts of the electro-magnetic theory.

SECTION IV. DIFFRACTION

174. Returning now to the equation of Art. 170, we know that if

$$\ddot{\xi} = c^2\nabla^2\xi,$$

we may put

$$\xi = \cosh(ct\nabla)\chi + \frac{\sinh(ct\nabla)}{ct\nabla}\psi \quad \text{............(1)},$$

χ, ψ being arbitrary functions of x, y, z.

Suppose, initially,

$$\dot{\xi} = F, \quad \xi = f \quad \text{........................(2)},$$

where F, f are given functions of x, y, z.

Then the solution of the above equation may be put in the form

$$\xi = \cosh(ct\nabla) f + \frac{\sinh(ct\nabla)}{ct\nabla} F \quad \text{............(3)},$$

where the arbitrary functions are replaced by known functions. For, obviously,

$$\xi_0 = \chi = f \text{ and } \dot{\xi}_0 = \frac{1}{c\nabla} F$$

(putting $t = 0$ in the solution).

In order to interpret this symbolic solution, let (α, β, γ) be a point (P) on a sphere of radius r, centre $O(x, y, z)$, and consider the integral

$$\int e^{\left(\alpha\frac{\partial}{\partial x} + \beta\frac{\partial}{\partial y} + \gamma\frac{\partial}{\partial z}\right)} dS,$$

where dS is an element of surface of the sphere.

On transformation, this can obviously be written

$$\int e^{z\frac{\partial}{\partial z}} dS;$$

so that

$$\frac{\partial^2}{\partial z^2} \equiv \nabla^2.$$

If now we put $z = r\cos\theta$, $dS = 2\pi r^2 \sin\theta\, d\theta$, the integral becomes

$$2\pi r^2 \int_0^\pi e^{r\cos\theta\nabla} \sin\theta\, d\theta = 4\pi r^2 \frac{\sinh r\nabla}{r\nabla}.$$

Putting now $r = ct$, and remembering that $dS = r^2 d\Omega$, where $d\Omega$ is the solid angle, subtended by dS at the centre,

$$\frac{\sinh ct\nabla}{c\nabla} = \frac{t}{4\pi}\int (e^{\alpha\frac{\partial}{\partial x} + \dots + \dots})\, d\Omega;$$

$$\therefore \quad \frac{\sinh ct\nabla}{c\nabla} F = \frac{t}{4\pi}\int (e^{\alpha\frac{\partial}{\partial x} + \dots + \dots})\, F d\Omega$$

$$= \frac{t}{4\pi}\int F(x+\alpha,\ y+\beta,\ z+\gamma)\, d\Omega.$$

Writing now $\alpha = lct$, etc., we have the symbolic solution replaced by

$$\xi = \frac{t}{4\pi}\int F(x+lct,\ \text{etc.})\, d\Omega + \frac{1}{4\pi}\frac{d}{dt}\left[t\int f(x+lct,\ \text{etc.})\right] d\Omega \quad \text{......(4).}$$

175. Hence, to find the effect at $O\,(x, y, z)$ at the time t, due to an initial disturbance f, F as defined in (2), we describe a sphere of radius ct, with O as centre, note the effects at all points over the sphere and take the sum as in the above expression.

In other words, if u_0 is the initial velocity, and ξ_0 is the initial displacement in the direction of x of any point on a sphere of radius ct, having O for origin, then the displacement at O at any time t, in the direction of x, can be briefly written from (4)

$$= \frac{t}{4\pi}\int (u_0)\, d\Omega + \frac{1}{4\pi}\frac{d}{dt} t \int (\xi_0)\, d\Omega \quad \text{............(5)}$$

which is Poisson's result.*

176. Thus, the expression for the disturbance, $[f(t)]_O$, at a point O, due to an element subtending $d\Omega$ at O, can be written from (5)

$$= \frac{t-t'}{4\pi} \cdot U_{t'} d\Omega + \frac{1}{4\pi}\frac{d}{d(t-t')}[(t-t') f t']\, d\Omega \text{......(6),}$$

where $U_{t'}$ is the resultant velocity at time t' at the sphere centre O and of radius $c(t-t')$; and the above construction shows that, if we put

$$t - t' = \frac{r}{c},$$

then
$$d\Omega = \frac{dS}{r^2}\cos nr,$$

* Rayleigh's *Sound*, Vol. II.

and
$$U_{t'} = \frac{\partial}{\partial t} f\left(t - \frac{r}{c}\right)$$
$$= -c\frac{\partial}{\partial r} f\left(t - \frac{r}{c}\right)$$
$$= \frac{-c}{\cos nr} \cdot \frac{\partial}{\partial n} f\left(t - \frac{r}{c}\right)$$

[$\cos nr$ = angle between r and the normal, drawn outwards];

$$\therefore \quad \frac{t-t'}{4\pi} U_{t'} d\Omega = -\frac{r}{4\pi} \frac{\partial}{\partial n} \left\{ f\left(t - \frac{r}{c}\right) \right\} \frac{dS}{r^2} \quad \text{......(7);}$$

$$\therefore 4\pi [f(t)]_0 = \left[\frac{\partial}{\partial r} \left\{ \frac{1}{r} f\left(t - \frac{r}{c}\right) \right\} \cos nr - \frac{1}{r} \frac{\partial}{\partial n} f\left(t - \frac{r}{c}\right) \right] dS \quad \text{..(8),}$$

which is Kirchhoff's solution.

177. In particular, if r_1 is the distance of the centre of a diverging spherical wave from dS, and .

$$f(t) \text{ (at } dS) = \frac{A}{r_1} \cos\left(\frac{2\pi t}{T} - \frac{r_1}{\lambda}\right)^*,$$

then it can be shown that at any point Q (whose distance from dS is r) the disturbance is

$$\frac{A}{2\lambda} \int \frac{1}{rr_1} \sin 2\pi \left(\frac{t}{T} - \frac{r + r_1}{\lambda}\right) (\cos nr - \cos nr_1)\, dS. \quad \text{[Drude.]}$$

178. Remembering that the direction of displacement must be always perpendicular to the direction of propagation, the above result will be correct only on the supposition that the direction of vibration is perpendicular to both r and r_1.

If this is not the case, we must resolve the displacement (which is necessarily perpendicular to r_1) in the direction perpendicular to r (in the plane of r and the vibration). In doing so, we observe that if u', v', w' be the components of the velocity perpendicular to r, and u, v, w those perpendicular to r_1, then $u' = u - ql$, etc.,
where q is the component of u, v, w along r.

* In view of the fact that the disturbance (s) from a point-source has to satisfy the differential equation

$$\frac{d^2(rs)}{dt^2} = \left(\frac{\lambda}{T}\right)^2 \cdot \frac{d^2}{dr^2}(rs).$$

This is derived from $c^2\nabla^2 s = \ddot{s}$, by transforming the equation into polars and retaining only the r-term and putting $cT = \lambda$.

In Poisson's solution, then, u_0 must be changed into $u_0 - q_0 l$, a corresponding change being effected in the second term.

The modified equations will then be (neglecting terms in r^{-2})

$$4\pi\xi = \frac{u_0 - q_0 l}{r}\, dS + \left(\frac{\partial \xi_0}{\partial r} - l\frac{\partial \rho_0}{\partial r}\right)\frac{dS}{r}, \text{ etc.} \quad \ldots\ldots(9),$$

where $$\rho_0 = l\xi_0 + m\eta_0 + n\zeta_0,$$

(l, m, n being the direction-cosines of r), and, ξ, η, ζ, the required displacements due to ξ_0, η_0, ζ_0 at dS.

Taking now the particular case of a displacement along the axis of z, viz. $f\left(t - \frac{x}{c}\right)$ (the direction of propagation being along the axis of x), we have, neglecting terms depending on r^{-2}, and remembering that the resolved part of the displacement perpendicular to r is $f(t')\sin\phi$, where ϕ is the angle made by r with the axis of z, from (8) [Art. 176],

$$4\pi\,[f(t)]_o = \frac{\sin\phi}{cr}\left[f'\left(t - \frac{r}{c}\right)(1 - \cos nr)\right] dS$$

$$= \frac{\sin\phi}{cr} f'\left(t - \frac{r}{c}\right)(1 + \cos\theta)\, dS,$$

if θ is the angle between r and the direction of propagation, both being measured in the same direction.

In particular, if

$$f\left(t - \frac{x}{c}\right) = A\sin\frac{2\pi c}{\lambda}\left(t - \frac{x}{c}\right)$$

$$[f(t)]_o = \frac{1}{4\pi}\frac{A}{c}\cdot\frac{2\pi c}{\lambda}\cdot\frac{1}{r}\left[\cos\frac{2\pi c}{\lambda}\left(t - \frac{r}{c}\right)\right]\sin\phi\,(1 + \cos\theta)\, dS$$

$$= \frac{A}{2\lambda r}\left[\cos\frac{2\pi c}{\lambda}\left(t - \frac{r}{c}\right)\right]\sin\phi\,(1 + \cos\theta)\, dS \quad \ldots\ldots(10).$$

(9) and (10) are Stokes's results.

179. Now, according to Rowland, electric and magnetic displacements being perpendicular to each other, and the mean energies being equal, the total illumination must vary as $1 + \cos\theta$.

He further argues that Stokes's solution is based on displacement and rate of displacement of the elastic medium, but in an elastic medium, there is not only displacement but rotation also, and the components of this rotation must also satisfy the equation of continuity. But when a wave is broken up at an orifice, the rotation is left discontinuous by Stokes's solution. Now, as the equation of propagation of a rotation is the same as that of a displacement, and the two are at right angles to each other, both are important.

Hence, according to Rowland, on the elastic solid theory as well as on the electro-magnetic theory, the true solution of diffraction will depend on the sum of two similar terms; and it will follow, therefore, that diffraction cannot supply the criterion, whereby we may determine the relation between displacement and polarisation.

Glazebrook has, however, pointed out that the magnetic displacement is in consequence of the electric, and if we take account of the latter, we have done all that is necessary. Rowland's results, moreover, contradict the results experimentally verified and theoretically obtained by Lord Rayleigh on the blue of the sky.

180. In any case, for the production of diffraction effect we must have $\phi = \frac{\pi}{2}$ nearly. On this understanding we can work out the case of oblique incidence at a narrow slit as follows:

Let x be the distance of a small element dx from the centre of the slit. The disturbance reaching any point will be [since the difference of phase due to disturbance from the central element and that at a distance x is equal to $\frac{x}{\lambda}(\sin i - \sin\theta)$]

$$= \frac{\cos i + \cos(i-\theta)}{2\lambda r}\int_{-\frac{l}{2}}^{\frac{l}{2}} dx \,.\, \sin 2\pi\left(\frac{t}{T} - \frac{\delta}{\lambda} - \frac{x}{\lambda}(\sin i - \sin\theta)\right)$$

$$= \frac{1}{4\pi}\,\frac{\cos i + \cos(i-\theta)}{\sin i - \sin\theta}\sin\left(\frac{l}{\lambda}(\sin i - \sin\theta)\right).\sin 2\pi\left(\frac{t}{T} - \frac{\delta}{\lambda}\right),$$

where $2\pi\left(\frac{t}{T} - \frac{\delta}{\lambda}\right)$ is the incident disturbance, and l is the breadth of the slit.

On either side of the zero position, the disturbance will therefore be

$$\frac{\cos i + \cos (i \pm \theta)}{\sin i \pm \sin \theta} \sin \left(\frac{l}{\lambda} (\sin i \pm \sin \theta)\right),$$

which accounts for the want of symmetry observed by Mr C. V. Raman* when a diffraction band is produced at oblique incidence.

Section V. Reflection and Refraction

181. The equations of Maxwell are of the type:

$$u = \dot{f} + p \quad \text{.........................(1)},$$

$$f = \frac{K}{4\pi} X \quad \text{.........................(2)},$$

$$p = \sigma X \quad \text{.........................(3)},$$

$$X = -\dot{F} - \frac{\partial \psi}{\partial x} \quad \text{.........................(4)},$$

$$a = \frac{\partial H}{\partial y} - \frac{\partial G}{\partial z} \quad \text{.........................(5)}$$

$$= \mu \alpha \quad \text{.........................(6)},$$

$$4\pi u = \frac{\partial \gamma}{\partial y} - \frac{\partial \beta}{\partial z} \quad \text{.........................(7)}.$$

Moreover, the conditions to be satisfied at the surface of separation of two media are the continuity of tangential magnetic forces and of normal induction—magnetic as well as electric.

From (1), (6), (7),

$$4\pi (p + \dot{f}) = \frac{\partial}{\partial y}\left(\frac{c}{\mu}\right) - \frac{\partial}{\partial z}\left(\frac{b}{\mu}\right)$$

$$= 4\pi \left(\dot{f} + \frac{4\pi\sigma}{K} f\right) \text{ from (2) and (3)} \quad \text{...(8)},$$

Again from (2), (4), (5),

$$4\pi \left[\frac{\partial}{\partial y}\left(\frac{f}{K}\right) - \frac{\partial}{\partial x}\left(\frac{g}{K}\right)\right] = \dot{c} \text{..............(9)},$$

* *Phil. Mag.* Jan. 1909.

therefore from (8) and equations of the type (9) we have, eliminating a, b, c,

$$\left(\ddot{f}+\frac{4\pi\sigma}{K}\dot{f}\right)=\left[\frac{\partial}{\partial x}\left(\frac{1}{\mu}\frac{\partial}{\partial x}\frac{f}{K}\right)+\frac{\partial}{\partial y}\left(\frac{1}{\mu}\frac{\partial}{\partial y}\frac{f}{K}\right)\right.$$
$$\left.+\frac{\partial}{\partial z}\left(\frac{1}{\mu}\frac{\partial}{\partial z}\frac{f}{K}\right)\right]-\frac{\partial}{\partial x}\left[\frac{1}{\mu}\left\{\frac{\partial}{\partial x}\frac{f}{K}+\frac{\partial}{\partial y}\frac{g}{K}+\frac{\partial}{\partial z}\frac{h}{K}\right\}\right].$$

If μ and K are constants, we have

$$\left(\ddot{f}+\frac{4\pi\sigma}{K}\dot{f}\right)=\frac{1}{\mu K}\nabla^2 f-\frac{\partial}{\partial x}\left(\frac{\partial f}{\partial x}+\ldots+\ldots\right);$$

similarly, $\left(\ddot{a}+\frac{4\pi\sigma}{K}\dot{a}\right)=\frac{1}{\mu K}\nabla^2 a-\frac{\partial}{\partial x}\left(\frac{\partial a}{\partial x}+\ldots+\ldots\right).$

That is, generally, $\ddot{\phi}=V^2\nabla^2\phi$ if $\sigma=0$ and $\mu K V^2=1$.

182. Case I. Let $f=0=g$ at $z=0$; that is, let electric displacement be perpendicular to the plane of incidence ($z=0$). Let, also, $\phi\equiv\frac{h}{K}$, and let $x=0$ be the refracting surface.

Then ϕ and $\frac{1}{\mu}\frac{\partial\phi}{\partial x}$ must be continuous, at $x=0$.

Let $\phi_1=a_1\sin\frac{2\pi}{\lambda}(Vt-\overline{lx+my})$, incident wave,

$\phi_2=a_2\sin\frac{2\pi}{\lambda}(Vt-\overline{my-lx})$, reflected wave,

$\phi_3=a_3\sin\frac{2\pi}{\lambda}(V't-\overline{l'x+m'y})$, refracted wave.

The continuity of ϕ yields $a_1+a_2=a_3$ and $\frac{m}{m'}=\frac{\lambda}{\lambda'}=\frac{V}{V'}$, and the continuity of $\frac{1}{\mu}\frac{\partial\phi}{\partial x}$ yields $\frac{a_1-a_2}{\mu_1}\frac{l}{\lambda}=\frac{a_3}{\mu_2}\cdot\frac{l'}{\lambda'}$,

i.e. $$\frac{a_1-a_2}{a_1+a_2}=\frac{\mu_2}{\mu_1}\cdot\frac{\tan r}{\tan i}\text{, as in Art. 84,}$$

and $$a_3=a_1\frac{2\sqrt{\mu_1\mu_2}\sin i}{(\mu_2\tan r+\mu_1\tan i)\cos r}.$$

183. Case II. $h=0=a=b$.

Here, let $\phi=\frac{c}{\mu}$. Then the surface conditions are the continuity of ϕ and $\frac{1}{K}\frac{\partial\phi}{\partial x}$. And the results are obtained by changing μ into K in the above.

184. If we put $\mu_1 = \mu_2$, Case I gives Fresnel's formulae for reflection and refraction for light polarised in the plane of incidence. Comparing this with the assumption made there, we conclude that the electric displacement is perpendicular to the plane of polarisation, if Fresnel's results are to be accepted and if $\mu_1 = \mu_2$. Similarly, Case II yields formulae which are also Fresnel's for light polarised perpendicularly to the plane of incidence.

185. It follows, therefore, since the plane of incidence is the plane of polarisation, that the light vector is perpendicular to the plane of polarisation, provided the electric force is identified with that vector, and $\mu_1 = \mu_2$.

If, on the other hand, the light vector is identified with magnetic force, it will be in the plane of polarisation.

186. On applying the theory to the case in which K and μ are both variable, it is found that in order to account for the fact that the light scattered by small particles on which polarised light impinges does vanish in one direction perpendicular to the original ray, we must make* either ΔK or $\Delta\mu$ vanish, so that both cannot be variable.

This corresponds to the conclusion that both inertia and rigidity of the ether, supposed to be an elastic medium, cannot vary. Electrical evidence is in favour of the variation of K.

We have not equally certain experimental evidence as to the choice between rigidity and inertia. Both labile-ether theory and Green's theory necessitate the non-variation of rigidity.

187. For metallic reflection, $\sigma \neq 0$ in the second metallic medium, and therefore in the second medium K is to be changed into

$$K(1 - 2iV^{-1}\sigma\lambda K^{-1}).$$

When this is done, it is found that we obtain Cauchy's expression, but at the same time the experimental results obtained by Jamin cannot be reconciled with theory without supposing $K_1 : K_2$, that is, the real part of the square of the complex refractive index negative.

* Rayleigh, *Phil. Mag.* 1881, p. 89.

Maxwell's theory, therefore, requires modification in order to account for metallic reflection. [J J. Thomson, *Recent Researches*, Art. 356.]

188. The theory, however, is capable of fairly satisfactorily explaining most of the phenomena of optics. But the outstanding questions remain—what is the intimate nature of the medium that is the seat of electro-magnetic phenomena and what is the intimate nature of electricity and magnetism, of electric displacement and electric and magnetic stress? Optical effects are certainly due to periodic changes of some properties of a medium, which we call the ether. The direction of propagation of this change is the direction of the flow of energy in the medium. But of the intimate nature of the mechanism by which this energy is produced or redistributed and propagated, we have no knowledge. Various attempts have been made to push the analysis of phenomena beyond this limit. It would be interesting to consider some of these.

SECTION VI CONCORDANCE BETWEEN THE ELECTRO-MAGNETIC AND ELASTIC SOLID THEORIES

189. It will be useful at this stage to compare the unmodified results of the electro-magnetic and elastic solid theories.

If T is the kinetic energy of a strained elastic medium of density σ, having displacements ξ, η, ζ at a point (x, y, z), then

$$2T = \int \sigma (\dot{\xi}^2 + \dot{\eta}^2 + \dot{\zeta}^2)\, dx\, dy\, dz \ldots\ldots\ldots\ldots\ldots (1).$$

Again, if $W =$ the potential energy of deformation, then, on MacCullagh's theory, as well as Lord Kelvin's,

$$\delta W = \delta \int 2n (\omega_x^2 + \omega_y^2 + \omega_z^2)\, dx\, dy\, dz \quad \ldots\ldots\ldots (2),$$

where ω_x, etc. are molecular rotations, and δ the operator of the calculus of variation (n being the rigidity).

On the other hand, the electro-magnetic energy in a medium of permeability μ is

$$\frac{1}{8\pi} \int \mu (\alpha^2 + \beta^2 + \gamma^2)\, dx\, dy\, dz \ldots\ldots\ldots\ldots\ldots (3),$$

where α, β, γ are the components of magnetic force, while the electrostatic energy is

$$\int \frac{2\pi}{K}(f^2+g^2+h^2)\,dx\,dy\,dz \quad \text{...............(4)},$$

(f, g, h) being electric polarisation and K the specific inductive capacity.

190. On the mode of identification suggested by Larmor, we may, therefore, take

$$(\alpha, \beta, \gamma) \text{ proportional to } (\dot{\xi}, \dot{\eta}, \dot{\zeta}) \equiv \frac{d}{dt}(\xi, \eta, \zeta)$$

and $\qquad (f, g, h) \quad " \quad " \quad (\omega_x, \omega_y, \omega_z)$,

the constants being suitably adjusted so that

$$\frac{2\pi f}{\alpha} = \frac{\omega_x}{\xi}, \text{ etc.,}$$

which implies

$$\frac{1}{\sqrt{K\mu}} = \sqrt{\frac{n}{\sigma}} \quad \text{..........................(5)},$$

or the velocity $\left(=\frac{1}{\sqrt{K\mu}}\right)$ of electro-magnetic disturbance is equal to $\sqrt{\frac{n}{\sigma}}$ = that of the rotational wave in the elastic medium.

This also satisfies the usual relations in elasticity and electro-magnetism, since

$$2\omega_x = \frac{\partial \zeta}{\partial y} - \frac{\partial \eta}{\partial z},$$

and

$$4\pi f \propto \left(\frac{\partial \dot{\zeta}}{\partial y} - \frac{\partial \dot{\eta}}{\partial z}\right)$$

$$= \left(\frac{\partial \gamma}{\partial y} - \frac{\partial \beta}{\partial z}\right),$$

and further implies the condition

$$\frac{\partial f}{\partial x} + \ldots + \ldots = 0,$$

since

$$\frac{\partial \omega_x}{\partial x} + \frac{\partial \omega_y}{\partial y} + \frac{\partial \omega_z}{\partial z} = 0.$$

191. Further, the equation of small motion in an elastic medium (n, σ) is

$$(\kappa + \tfrac{4}{3}n)\frac{\partial\Delta}{\partial x} - 2n\left(\frac{\partial\omega_z}{\partial y} - \frac{\partial\omega_y}{\partial z}\right) = \sigma\ddot{\xi} \quad \text{...........(6)},$$

where κ is the modulus of compression, and Δ the cubical dilatation.

If $\Delta = 0$, we have, putting $\omega_z \equiv 2\pi h$, $\dot{\xi} \equiv \alpha$, and $\dfrac{n}{\sigma} = V^2$,

$$\dot{\alpha} = -4\pi V^2\left(\frac{\partial h}{\partial y} - \frac{\partial g}{\partial z}\right) \quad \text{..................(7)},$$

which is the same equation as is derivable from the theorem

$$-\frac{d}{dt}\int(la + mb + nc)\,dS = \int(X\,dx + Y\,dy + Z\,dz), \text{ [Art. 163]}$$

i.e., the line integral of E.M.F. is equal to the rate of decrease of lines of force embraced by a circuit*, and necessarily involves the supposition that

$$\frac{\partial\alpha}{\partial x} + \ldots + \ldots = 0.$$

192. Now applying the condition $\delta\int(T - W)\,dt = 0$, we get from (1) and (2)

$$\ddot{\omega}_x = V^2\nabla^2\omega_x \quad \text{........................(8)},$$

where V is the velocity of propagation. Thus, we derive at once

$$\ddot{f} = V^2\nabla^2 f, \text{ and two similar equations,}$$

in a medium defined by

$$\frac{\partial f}{\partial x} + \frac{\partial g}{\partial y} + \frac{\partial h}{\partial z} = 0.$$

193. If, however, the medium is defined, owing to the presence of electrons [Ch. IV], by

$$\frac{\partial f}{\partial x} + \frac{\partial g}{\partial y} + \frac{\partial h}{\partial z} = \rho \quad \text{......................(9)},$$

where ρ is the volume density of electricity of polarisation; and moreover, if

$$\frac{\partial\beta}{\partial x} - \frac{\partial\alpha}{\partial y} = 4\pi(\dot{h} + \rho w) \quad \text{..................(10)},$$

* Maxwell, Vol. II, end of Chapter VIII, note.

so that the total current is taken to consist of polarisation and convection (electronic) currents, (u, v, w) being the velocity of electrons, we must take

$$\dot{\omega}_z' = 2\pi(\dot{h} + \rho w), \text{ etc. } \quad\quad\text{(11)};$$

while, under statical conditions

$$\omega_z' = 2\pi h + \tfrac{1}{2}K\frac{\partial\phi}{\partial z}, \text{ etc. } \quad\quad\text{(12)}.$$

From (9) and (12) we get

$$K\nabla^2\phi + 4\pi\rho = 0 \quad\quad\text{(13)},$$

when the electrons are at rest; while from (9) and (11) we have, since $\frac{\partial\dot{\omega}_x'}{\partial x} + \ldots + \ldots = 0$,

$$\frac{\partial\rho}{\partial t} + \frac{\partial(\rho u)}{\partial x} + \ldots + \ldots = 0,$$

which is the equation of continuity of a fluid of density ρ, and is also the condition postulated in the electron theory of Lorentz.

Now (7) gives, if we introduce the vector-potentials F, G, H, given as usual by

$$\mu\alpha = \frac{\partial H}{\partial y} - \frac{\partial G}{\partial z}, \text{ etc.}$$

$$-\dot{F} - \frac{\partial\psi}{\partial x} = \frac{4\pi f}{K}, \left[\text{where } \nabla^2\psi = 0 = \frac{\partial F}{\partial x} + \frac{\partial G}{\partial y} + \frac{\partial H}{\partial z}\right].$$

$$\therefore -\left(\dot{F} + \frac{\partial\psi}{\partial x}\right) \propto (\omega_x),$$

where ω_x is the molecular rotation in free ether.

This may be interpreted by saying that the force tending to produce free rotational displacement of the ether is $-\dot{F} - \frac{\partial\psi}{\partial x}$, etc.; while (12) shows that the force producing the total effect is proportional to $f + \frac{K}{4\pi}\frac{\partial\phi}{\partial x}$, etc. (a conclusion otherwise arrived at by Larmor, *loc. cit.*).

194. From (7) and (10) we have

$$\begin{aligned}\ddot{\alpha} &= -4\pi V^2\left[\frac{\partial \dot{h}}{\partial y} - \frac{\partial \dot{g}}{\partial z}\right]\\ &= V^2\left[\nabla^2\alpha - \frac{\partial}{\partial x}\left(\frac{\partial\alpha}{\partial x} + \frac{\partial\beta}{\partial y} + \frac{\partial\gamma}{\partial z}\right) + \frac{\partial(\rho w)}{\partial y} - \frac{\partial(\rho v)}{\partial z}\right]\\ &= V^2\left\{\nabla^2\alpha + \frac{\partial(\rho w)}{\partial y} - \frac{\partial(\rho v)}{\partial z}\right\},\end{aligned}$$

while, also from (7) and (10) we get, since

$$\frac{\partial f}{\partial x}+\frac{\partial g}{\partial y}+\frac{\partial h}{\partial z}=\rho,\ \ddot{h}-V^2\nabla^2 h=-\ V^2\frac{\partial \rho}{\partial z}-\frac{d}{dt}(\rho w).$$

Thus, on the whole, the motion in the general case depends on an equation of the form

$$\left(\frac{d^2}{dt^2}-c^2\nabla^2\right)\phi=F(x,\ y,\ z,\ t).\quad \text{[Lorentz.]}$$

195. In order to solve*

$$\left(\frac{d^2}{dt^2}-c^2\nabla^2\right)\phi=F\,(x,\ y,\ z,\ t)$$

we have only to find the particular integral†.

Proceeding in the usual way,

$$\phi=\frac{1}{D^2-c^2\nabla^2}F,\ \text{where}\ D\equiv\frac{d}{dt},$$

$$=\frac{1}{2D}\left[\frac{1}{D+c\nabla}+\frac{1}{D-c\nabla}\right]F$$

$$=\frac{1}{2D}\left[e^{-ct\nabla}\int e^{-ct'\nabla}\,F'\,dt'+e^{ct\nabla}\int e^{-ct'\nabla}\,F'\,dt'\right]$$

$$=\frac{1}{D}\int\{\cosh c\,(t-t')\,\nabla\}\,F'\,dt'=\frac{1}{D}\int\{\cosh c\,(t-t')\,\nabla\}\,Fdt$$

$$=\frac{1}{D}\frac{1}{4\pi}\int\left[\frac{d}{dt}\int(t-t').\,F\,\{x+c\,(t-t'),\ldots\}\,d\Omega\right]dt,$$

attending to the meaning of the operation $\cosh\,\{c\,(t-t')\,\nabla F\}$ [Art. 174],

$$=\frac{1}{4\pi}\iint r\,.\,F\left(t-\frac{r}{c}\right)d\Omega\,\frac{dr}{c^2},\ \text{since}\ c\,(t-t')=r\ \text{[Art. 176]},$$

$$=\frac{1}{4\pi c^2}\int\frac{F\left(t-\dfrac{r}{c}\right)}{r}\,d\tau,$$

where $d\tau$ is an element of volume, which is Lorentz's result.

It thus appears that complete solution depends on $t-\dfrac{r}{c}$. The physical meaning of this is worthy of note. For it shows that effect at time t depends on a time $t'\left(=t-\dfrac{r}{c}\right)$. t may then be called the local time and t', the standard time.

* *Phil. Mag.* Aug. 1914.

† Remembering that the *General* solution is given by Poisson's result.

196. Consider, now, a medium for which $K\ (k_x, k_y, k_z)$ depends on direction. W is then of the form

$$2\pi \int (k_x^{-1} f^2 + k_y^{-1} g^2 + k_z^{-1} h^2)\, dx\, dy\, dz$$

$$= \frac{1}{2\pi} \int (k_x^{-1} \varpi_x^2 + k_y^{-1} \varpi_y^2 + k_z^{-1} \varpi_z^2)\, dx\, dy\, dz,$$

if we take $2\pi f = \varpi_x$, etc.

Hence as on MacCullagh's theory, applying

$$\delta \int (T - W)\, dt = 0,$$

where $T = \frac{\mu}{8\pi} \int (\alpha^2 + \beta^2 + \gamma^2)\, dx\, dy\, dz$

$$= \frac{\mu}{8\pi} \int (\dot{\xi}^2 + \dot{\eta}^2 + \dot{\zeta}^2)\, dx\, dy\, dz, \text{ if } \alpha \equiv \dot{\xi}, \text{ etc.},$$

we get
$$\mu \ddot{\xi} = 2k_y^{-1} \frac{\partial \varpi_y}{\partial z} - 2k_z^{-1} \frac{\partial \varpi_z}{\partial y}.$$

But
$$2\dot{\varpi}_x = \frac{\partial \dot{\zeta}}{\partial y} - \frac{\partial \dot{\eta}}{\partial z} = 4\pi \dot{f}.$$

$$\therefore \quad \mu \ddot{f} = k_x^{-1} \nabla^2 f - \frac{\partial}{\partial x} \left(k_x^{-1} \frac{\partial f}{\partial x} + k_y^{-1} \frac{\partial g}{\partial y} + k_z^{-1} \frac{\partial h}{\partial z} \right)$$

giving Fresnel's wave surface, f, g, h being in the wave-front and perpendicular to the plane of polarisation. This result can, of course, also be obtained by direct application of equations of the electro-magnetic field.

Thus, we have

$$f = \frac{k_x}{4\pi} X,$$

and
$$4\pi\mu \dot{f} = \frac{\partial c}{\partial y} - \frac{\partial b}{\partial z},$$

i.e.,
$$4\pi\mu \ddot{f} = \left(\frac{\partial}{\partial y} \dot{c} - \frac{\partial}{\partial z} \dot{b} \right),$$

the medium being stationary.

$$\therefore \quad 4\pi\mu \ddot{f} = \frac{\partial}{\partial y} \left(\frac{\partial X}{\partial y} - \frac{\partial Y}{\partial x} \right) - \frac{\partial}{\partial x} \left(\frac{\partial Z}{\partial x} - \frac{\partial X}{\partial z} \right). \quad \text{[Art. 163.]}$$

$$\therefore \quad \mu \ddot{f} = \nabla^2 (k_x^{-1} f) - \frac{\partial}{\partial x} \left[\frac{\partial}{\partial x} k_x^{-1} f + \frac{\partial}{\partial y} k_y^{-1} g + \frac{\partial}{\partial z} k_z^{-1} h \right].$$

Hence, ultimately,

$$\mu \ddot{f} = k_x^{-1} \nabla^2 f - \frac{\partial}{\partial x}\left(k_x^{-1}\frac{\partial f}{\partial x} + k_y^{-1}\frac{\partial g}{\partial y} + k_z^{-1}\frac{\partial h}{\partial z}\right),$$

assuming k_x, k_y, k_z to be absolute constants, depending on the directions of the axes only.

197. The agreement of the results might at first sight seem to indicate a means of physically interpreting the electric and magnetic quantities, in terms of the elastic and inertia constants of the medium. But an initial difficulty presents itself. μ is practically constant for all (non-magnetic) media, but σ is most probably not, and while n (the rigidity) is most probably constant for all media, K is not. Moreover, the same agreement can be reached by other modes of identification. [Cf. molecular vortex theory.] In fact, the fundamental assumption, on which the concordance is based, itself seems not to be altogether legitimate. For we have no means of deciding which of the two forms of energy in which the total energy of an electro-magnetic field or of the elastic medium is assumed to exist is kinetic, and which potential. It seems rather to indicate that the distinction between potential and kinetic energy is more conventional than real. It is, of course, conceivable that $\frac{n}{\sigma} = \frac{1}{\mu K}$, though we may not be able to discover how the several quantities are separately related to each other.

198. Comparing, now, the equations of the electro-magnetic theory with those appropriate to the various elastic solid theories, we deduce the following formal relations*.

On the electro-magnetic theory, electric and magnetic inductions, being circuital, are in the plane of the wave front but electric and magnetic forces are not necessarily so (except in optically homogeneous media). A complete agreement is arrived at by making $\mu = 1$ in all cases as on the electron theory.

Electric induction or polarisation is to be regarded as

* Larmor, *Philosophical Transactions*, 1894, Part II.

identical with mechanical displacement, *as conceived by Fresnel.*

Magnetic induction is the same as the mechanical displacement of MacCullagh's theory.

Electric or magnetic force is the same as mechanical displacement on the theories of Boussinesq and Kelvin, respectively.

Further, æleotropic inertia combined with labile isotropic elasticity yields Fresnel's wave surface and MacCullagh's theory of crystalline reflection and refraction.

If the magnetic quality of the medium takes part in vibrations (μ, not constant) there will be an apparent alteration of density with medium, but this will not make the normal displacement continuous, for the normal induction must be necessarily so. For this, there should be suitable labile properties.

If magnetic quality is ineffective, the degree of compressibility will be immaterial.

Again, on the assumption that

(1) energy is propagated along rays,

(2) it is half potential, half kinetic,

(3) there is no loss of energy,

the solution of the problem of reflection, as distinct from elastic constitution of a medium in which light is polarised linearly, is characterised as follows:

If density is taken to be uniform, we get MacCullagh's theory on which vibrations are in the plane of polarisation. On this theory, the ether in crystalline media is possessed of æleotropic rotational elasticity and isotropic effective inertia (the same in all media), whatever be the compressibility.

If we regard the ether to be an isotropic solid but possessed of æleotropic effective inertia, and if, further, we assume that there is no elastic discontinuity, as we pass from one medium to another, that is, that the elasticity is the same in all media, then all the conditions are satisfied, if there is no resistance to laminar compression.

The condition of continuity of energy is of the same form in all cases.

SECTION VII. HYDRODYNAMIC ANALOGY

199. The following analogy between quantities in hydrodynamics and electro-magnetism may be noticed.

Let (u, v, w) be the velocity of a fluid and (α, β, γ) the magnetic force; then if we take $(u, v, w) \propto (\alpha, \beta, \gamma)$, since

$$\frac{\partial u}{\partial x} + \frac{\partial v}{\partial y} + \frac{\partial w}{\partial z} = \theta,$$

where θ is the expansion, $\theta \propto m$,

where $$\frac{\partial \alpha}{\partial x} + \ldots + \ldots = m,$$

and m is the volume density of imaginary magnetic matter.

If $(\dot{\varpi}_x, \dot{\varpi}_y, \dot{\varpi}_z)$ = vortex spin, (p, q, r) = electric current, then

$$2\dot{\varpi}_x = \frac{\partial w}{\partial y} - \frac{\partial v}{\partial z}, \text{ etc.,}$$

and this is consistent with

$$4\pi p = \frac{\partial \gamma}{\partial y} - \frac{\partial \beta}{\partial z}, \text{ etc.,}$$

if $(\dot{\varpi}_x, \dot{\varpi}_y, \dot{\varpi}_z) \propto (p, q, r)$.

Again, since (F, G, H) is defined by

$$u = \frac{\partial H}{\partial y} - \frac{\partial G}{\partial z}, \text{ in Hydrodynamics,}$$

while $$\mu\alpha = \frac{\partial H}{\partial y} - \frac{\partial G}{\partial z}, \text{ in Electro-magnetism,}$$

the analogy may be said to be complete.

We observe further that these equations include the previous mode of identification [Art. 191], viz. $\dot{\varpi}_x \propto f$ and $u \propto \dot{\xi}$.

200. If we admit that the above analogy is not merely formal but has a physical basis, we are led to some such interpretation of electro-magnetic phenomena as the following*.

A permanent magnetic element will be represented by a circuital cavity or channel in the (elastic) ether, along the surface of which there is a distribution of vorticity. It will, in short, be a vortex ring with a vacuum for its core.

Atoms will be ordinary coreless vortices, the velocity of the primordial fluid constituting magnetic force. When a piece of

* Larmor, "On a Dynamical Theory of the Electric and Luminiferous Medium."

matter is electrified, an elastic rotational displacement would be set up in the ether surrounding it, the absolute rotation at each point representing the electric displacement.

A charged body in the field causes a rotational strain in the ether all around it.

If the motion of the ether represents magnetic force, the fact that μ is practically the same in all sensibly non-magnetic bodies as in vacuum must be taken to indicate that the ether flows with practically its full velocity in all such media, so that there is very little obstruction interposed on its motion by the presence of matter.

201. As to magneto-optic rotation, we may readily admit (with Larmor) that

"the rotation of the plane of polarisation in a uniform magnetic field depends on the interaction of the uniform velocity of the ether which constitutes that field [on the above analogy] with the vibrational velocity of light disturbance.

For light waves, the motion that is elastically effective is the spin $(\dot{\varpi}_x, \dot{\varpi}_y, \dot{\varpi}_z) = (\dot{f}, \dot{g}, \dot{h})$, and the varying part of the velocity of an element of volume containing rotational motion of the magnetic vortices is proportional to

$$\left(\alpha_0 \frac{\partial}{\partial x} + \dots + \dots\right)(\xi, \eta, \zeta) \equiv \frac{\partial}{\partial \theta}(\xi, \eta, \zeta) \text{ (say)},$$

(ξ, η, ζ) being the displacement of the medium, and $(\alpha_0, \beta_0, \gamma_0)$ the impressed magnetic field. This variation is caused by alteration of the vibrational velocity of a particle owing to its change of position, as it is carried along in the magnetic field, as in hydrodynamics. There may thus exist a term in the energy, resulting from this interaction, of the form

$$C\left(\frac{\partial \xi}{\partial \theta}\dot{f} + \frac{\partial \eta}{\partial \theta}\dot{g} + \frac{\partial \zeta}{\partial \theta}\dot{h}\right), \text{ where } C \text{ is a constant.''}$$

202. To see this more clearly, assume that the kinetic energy of the medium under an impressed magnetic force contains a term (on hydrodynamic analogy) of the form

$$2C'(\alpha w_1 + \beta w_2 + \gamma w_3) \quad \dots\dots\dots\dots\dots\dots(1),$$

where w_1, w_2, w_3 are vortical spins of the medium, and $(\alpha, \beta, \gamma) = (u, v, w)$.

On the mode of identification here used, viz. $w_1 \propto \dot{f}$, etc., and since ξ, η, ζ are to be the displacements of the medium,

$$\frac{D\xi}{Dt} = \frac{\partial\xi}{\partial t} + \left(u_0 \frac{\partial}{\partial x} + \dots + \dots\right)\xi,$$

where $\frac{D}{Dt}$ is the total variation in time, the motion of the medium being taken into account; accordingly the effect of the impressed magnetic forces α_0, β_0, $\gamma_0 (\equiv u_0, v_0, w_0)$ is obtained by writing for α, in (1),

$$\left(\alpha_0 \frac{\partial}{\partial x} + \dots + \dots\right)\xi, \text{ etc.}$$

Hence the above expression (1) becomes

$$C\left(\frac{\partial\xi}{\partial\theta}\dot{f} + \dots + \dots\right).$$

Thence, as in Art. 210, we get the terms in the equation of motion appropriate to the Hall effect.

Thus, we observe that magnetic rotation can be explained independently of the theory that magnetic force is a rotation of the luminiferous medium.

203. Again, the velocity of light is not sensibly altered by motion along a field of electric displacement. Therefore, we cannot connect electric displacement with any considerable body-velocity of the ether. But the same difficulty attaches to the present theory, for Professor Lodge has found that the velocity of light is unaffected by even a very strong field of magnetic force. This leads to the conclusion, on either view, that ethereal density must be great.

204. In discussing the theory to which the above analogy leads, Larmor remarks as follows:

"The foundation of the present view is the conception of a medium which has the properties of a perfectly incompressible fluid, as regards irrotational motion, but is at the same time endowed with an elasticity which allows it to be the seat of energy of strain and to propagate undulations of a transverse type. This idea of a fluid medium disposes of the well-known difficulties which pressed on all theories that impose on the ether the quality of solidity. If a perfect fluid is a mathematical abstraction, the rotational elasticity with which the medium is endowed effectually prevents any slip or breach such as would be the point of failure of a simple fluid medium without some special quality to ensure continuity of motion.

"The objection has been taken to the idea of identifying velocity in hydrodynamics with magnetic force in electric phenomena, that it would be impossible on such a theory for a body to acquire a charge of electricity. A cardinal feature in the electrical development of the present theory is, on the other hand, the conception of intrinsic rotational strain constituting electric charge ($f = \varpi_x$, etc.), which can be associated with an atom or an electric conductor and which cannot be discharged without rupture of the continuity of the medium.... An atom is a vortex ring with associated intrinsic strain. The elastic effect of convection, through the medium of an atom thus charged, is equivalent to that of a twist round its line of movement. Such a twist is thus a physical element of an electric current."

205. It is noteworthy that we are able to derive the equations giving the fundamental assumption of the electron theory [Ch. IV] from the hypothesis of rotational strain [Art. 193].

SECTION VIII. MOLECULAR VORTEX THEORY

206. The consideration of the action of magnetism on polarised light would seem naturally to lead to the conclusion that, in a medium under the action of magnetic forces, something in the nature of an angular velocity, whose axis is in the direction of the magnetic force, forms part of the phenomena, corresponding to that of molecular rotation of an elastic solid or of angular velocity of vortex motion in hydrodynamics. If the latter analogy is to be complete, the displacement of the medium during the propagation of light should produce a disturbance of the vortices.

Pursuing this analogy, let us take

$\dot{\varpi}_x, \dot{\varpi}_y, \dot{\varpi}_z$ (vortex spin) $\equiv$ magnetic force (α, β, γ).

Then we have $$2\varpi_x = \frac{\partial w}{\partial y} - \frac{\partial v}{\partial z},$$

corresponding to $$\mu\alpha = \frac{\partial H}{\partial y} - \frac{\partial G}{\partial z} \quad \text{.....................(1)},$$

and therefore $$F = \tfrac{1}{2}\mu u, \text{ etc.} \quad \text{.....................(2)},$$

where u, v, w, F, G, H have their usual meaning in hydrodynamics and electro-magnetism; i.e. u, v, w are the velocities of the fluid medium, and F, G, H the components of electro-kinetic momentum; therefore the electro-kinetic momentum $= \frac{1}{2}\mu\,(u, v, w)$ = the momentum of the primordial fluid, provided we admit that $\mu \propto \rho$, where ρ is the density of the fluid.

This is the molecular vortex theory of Maxwell.

Moreover, since $$\frac{\partial\gamma}{\partial y} - \frac{\partial\beta}{\partial z} = 4\pi\dot{f},$$

$$-\nabla^2 F + \frac{\partial}{\partial x}\left(\frac{\partial F}{\partial x} + \ldots + \ldots\right) = 4\pi\mu\dot{f}, \text{ from (1)} \quad \text{...(3)},$$

μ being regarded as a constant.

207. If the primordial fluid is a viscous mass, the equations of motion will be

$$\rho\dot{u} = \rho X - \frac{\partial p}{\partial x} + \frac{1}{3}k\frac{\partial}{\partial x}\left(\frac{\partial u}{\partial x} + \frac{\partial v}{\partial y} + \frac{\partial w}{\partial z}\right) + k\nabla^2 u, \text{ etc.} \ldots(4)$$

(where p = the pressure and k = coefficient of viscosity).

But since $F = \frac{1}{2}\mu u$, etc., we have from (2), (3), (4)

$$\frac{2\rho}{\mu}\dot{F} = \rho X - \frac{\partial p}{\partial x} + \frac{8}{3}\frac{k}{\mu}\frac{\partial}{\partial x}\left(\frac{\partial F}{\partial x} + \ldots + \ldots\right) - 8\pi k\dot{f} \ldots(5).$$

Now, (5) is of the form

$$P = -\dot{F} - \frac{\partial\psi}{\partial x} \quad \text{.....................(6)},$$

if we write $$P = \frac{4\pi\mu k}{\rho}\dot{f}$$

$\left(\text{that is, the electric conductivity } (C) = \dfrac{\rho}{4\pi\mu k}\right)$,

and $$-\frac{\partial\psi}{\partial x} = \frac{\mu X}{2} - \frac{\mu}{2\rho}\frac{\partial p}{\partial x} + \frac{4}{3}\frac{k}{\rho}\cdot\frac{\partial}{\partial x}\left(\frac{\partial F}{\partial x} + \ldots + \ldots\right) \ldots(7),$$

or $$\psi = V + \frac{\mu}{2\rho}p - \frac{4}{3}\cdot\frac{k}{\rho}\left(\frac{\partial F}{\partial x} + \ldots + \ldots\right) \quad \text{......(8)},$$

where
$$\frac{\partial V}{\partial x} = \frac{\mu X}{2}.$$

Now since $\frac{\partial f}{\partial x} + \ldots + \ldots = 0$, and $\frac{\dot{f}}{C} = -\dot{F} - \frac{\partial \dot{\psi}}{\partial x}$, etc.,

$$-\frac{\partial}{\partial t}\left(\frac{\partial F}{\partial x} + \ldots + \ldots\right) - \nabla^2 \dot{\psi} = 0.$$

For a conductor, since $\dot{\psi}$ is evidently the potential of free electricity, $\nabla^2 \dot{\psi} = 0$. Let, now, $\frac{\partial F}{\partial x} + \ldots + \ldots \equiv J$. Then

$$\frac{dJ}{dt} = 0, \quad \therefore \quad J = \text{const.} = \lambda \text{ (say).}$$

But the equation of continuity gives

$$\frac{\partial \rho}{\partial t} + \frac{\partial (\rho u)}{\partial x} + \ldots + \ldots = 0,$$

or
$$\frac{D\rho}{Dt} + \frac{\rho}{2\mu} J = 0, \quad \therefore \quad \rho = \rho_0 e^{-\frac{2\lambda t}{\mu}}.$$

But this is meaningless, since ρ is to be the density of the medium. Therefore λ must be $= 0$, or

$$J = 0 = \frac{\partial u}{\partial x} + \ldots + \ldots = 0,$$

that is, the primordial fluid is incompressible.

Further from (3)

$$-\dot{f} = \frac{1}{4\pi\mu} \nabla^2 F = \frac{1}{8\pi} \nabla^2 u,$$

i.e. the electric current $= -\frac{1}{8\pi} \nabla^2$ (the velocity of the fluid), while the electric displacement $= -\frac{1}{8\pi} \nabla^2$ (the displacement of the medium)*.

208. When C is very small, the viscous fluid becomes an elastic solid and the equations of motion with the body forces X, Y, Z are

$$\rho \dot{u} = \rho X + A \frac{\partial \Delta}{\partial x} + n \nabla^2 \xi, \text{ etc.}$$

where
$$\Delta \equiv \frac{\partial \xi}{\partial x} + \frac{\partial \eta}{\partial y} + \frac{\partial \zeta}{\partial z}$$

and A, n are constants.

* Glazebrook, *Phil. Mag.* June, 1881.

The concordance of these equations with the electromagnetic equations of Maxwell has already been noticed, with the condition between the constants, given by

$$\frac{\mu n}{\rho} = K = \text{specific inductive capacity.}$$

209. Consider, next, the magneto-optic energy.

Assume that the kinetic energy of the medium under impressed magnetic forces contains a term of the form

$$2C(\alpha w_1 + \beta w_2 + \gamma w_3) \quad \ldots\ldots\ldots\ldots\ldots\ldots(1),$$

that is, a term of the form

$$C\int\left[\alpha\left(\frac{\partial w}{\partial y} - \frac{\partial v}{\partial z}\right) + \ldots\right] dx\,dy\,dz$$

$$= C\int\left[u\left(\frac{\partial \gamma}{\partial y} - \frac{\partial \beta}{\partial z}\right) + \ldots + \ldots\right] dx\,dy\,dz \quad \ldots\ldots(2),$$

provided the surface integral vanishes.

But if (ξ, η, ζ) be the displacement of a fluid medium, and the fluid is incompressible,

$$\varpi_x = \left\{(\varpi_x)_0 \frac{\partial}{\partial x} + (\varpi_y)_0 \frac{\partial}{\partial y} + (\varpi_z)_0 \frac{\partial}{\partial z}\right\}\xi, \text{ etc.}$$

$$\therefore \quad \alpha = \left(\alpha_0 \frac{\partial}{\partial x} + \ldots + \ldots\right)\xi \equiv \frac{\partial \xi}{\partial h} \text{ (say)}$$

Hence, the expression (1) becomes

$$C\left[u\frac{\partial}{\partial h}\left(\frac{\partial \zeta}{\partial y} - \frac{\partial \eta}{\partial z}\right) + \ldots + \ldots\right] = C\int \dot{\xi}\,\frac{\partial \varpi_x}{\partial h} + \ldots + \ldots, \text{ from (2)},$$

where $\dot{\varpi}_x \equiv w_1$, etc. [Maxwell's *Elect. and Mag.* Vol. II.]

This suffices to explain magnetic action of light—both Verdet's results as well as Hall's phenomena. But it requires that μ should vary as density,—that is, postulates uniform density in contradiction to the previous theories.

210. Applying $\delta\int(T - W)\,dt = 0$, since

$$\delta W = \delta\int 2n(\varpi_x^2 + \varpi_y^2 + \varpi_z^2)\,dx\,dy\,dz,$$

$$\rho\ddot{\xi} + 2C\frac{\partial}{\partial h}\left(\frac{\partial \dot{\zeta}}{\partial y} - \frac{\partial \dot{\eta}}{\partial z}\right) = \rho X + n\nabla^2\xi,$$

or
$$\dot{F} + \frac{2\mu C}{\rho}\frac{\partial w_1}{\partial h} = \rho\,\frac{X}{2} - 4\pi\,\frac{n\mu f}{\rho},$$

since
$$F = \frac{\mu u}{2} = \frac{\mu\dot{\xi}}{2} \text{ and } -f = \frac{1}{8\pi}\nabla^2\xi, \text{ etc.}$$

But
$$\frac{\partial w_1}{\partial h} = \left(\alpha_0\frac{\partial}{\partial x} + \ldots + \ldots\right) w_1$$
$$= \alpha_0\frac{\partial w_1}{\partial x} + \beta_0\frac{\partial w_2}{\partial x} + \gamma_0\frac{\partial w_3}{\partial x}$$
$$- \beta_0\left(\frac{\partial w_2}{\partial x} - \frac{\partial w_1}{\partial y}\right) - \gamma_0\left(\frac{\partial w_3}{\partial x} - \frac{\partial w_1}{\partial z}\right)$$
$$= \frac{\partial}{\partial x}(\alpha_0 w_1 + \beta_0 w_2 + \gamma_0 w_3) + 4\pi(\gamma_0\dot{g} - \beta_0\dot{h}).$$

$$\therefore\ P = \frac{4\pi f}{K} = \frac{4\pi n\mu}{\rho} f$$
$$= -\dot{F} - \frac{\partial\psi'}{\partial x} - \frac{8\pi\mu C}{\rho}(\gamma_0\dot{g} - \beta_0\dot{h}),$$

where
$$\psi' = \psi + \frac{2\mu C}{\rho}(\alpha_0 w_1 + \ldots + \ldots),$$

giving the additional E.M.F. due to the impressed magnetic field*.

SECTION IX. MOVING MEDIA. ELECTRO-MAGNETIC THEORY OF ABERRATION

211. So far, we have been attempting to analyse the intimate nature of the electro-magnetic field on the basis of Maxwell's theory.

As we have already seen, however, Maxwell's theory is not capable of giving an adequate account of phenomena which involve the interaction of matter and ether. Take aberration for example. In order to apply this theory, we must modify the equations to suit the case of a moving medium.

212. If a circuit or medium in which there is a flow of electricity is at rest, we have $X = -\dot{F} - \frac{\partial\psi}{\partial x}$. [Art. 163.]

* Glazebrook, *Phil. Mag.* June, 1881.

When the medium is in motion, we have to take account of this motion in the equation. Thus, if D expresses total increment

$$\int (Xdx + Ydy + Zdz) = \frac{D}{Dt}\int (Fdx + Gdy + Hdz)$$
$$\equiv \frac{D}{Dt}\int Jds \equiv \frac{DL}{Dt} = \frac{dL}{dt} + \frac{\partial L}{\partial t}\text{, say.}$$

Let an element of circuit AB come to the position $A'B'$. Now the surface integral over $A'B'BA$

$$= \int_{A'}^{B'} Jds - \int_{A}^{B} Jds - \int_{B}^{B'} Jds + \int_{A}^{A'} Jds,$$

or $$\iint \left[l\left(-\frac{\partial G}{\partial z} + \frac{\partial H}{\partial y}\right) + \ldots + \ldots \right] dS = \partial L_{AB} - \int_{B}^{B'} Jds + \int_{A}^{A'} Jds,$$

where L_{AB} = the line integral from A to B and ∂L is the part of the increment of L due to motion.

But $$dS = ABA'B' = Udt\, ds \sin \alpha,$$

where $$AB = ds,\ BB' = Udt,\ \angle BAA' = \angle \alpha.$$

$$\therefore \quad \iint (la + mb + nc)\, dS = \int dt \begin{vmatrix} dx & dy & dz \\ u & v & w \\ a & b & c \end{vmatrix}$$

since $$ds = (dx, dy, dz), \text{ and } U = (u, v, w)$$

But $$\Sigma \int_{A}^{A'} Jds = 0 = \Sigma \int_{B}^{B'} Jds,$$

for a complete circuit. Hence we have ultimately,

$$\frac{\partial L}{\partial t} = \int \begin{vmatrix} dx & dy & dz \\ u & v & w \\ a & b & c \end{vmatrix}$$

and $$X = -\dot{F} - \frac{\partial \psi}{\partial x} + (vc - bw), \text{ etc.}$$

213. We are now in a position to discuss the phenomena of aberration on Maxwell's theory.

The equations of Art. 181 will be modified as follows:

$$4\pi \left[\frac{\partial}{\partial y}\left(\frac{f}{K}\right) - \frac{\partial}{\partial x}\left(\frac{g}{K}\right) \right]$$
$$= -\frac{d}{dt}\left(\frac{\partial F}{\partial y} - \frac{\partial G}{\partial x}\right) + \frac{\partial}{\partial y}(v_y c - b v_z) - \frac{\partial}{\partial x}(v_z a - c v_x),$$

(v_x, v_y, v_z) being the velocity of the medium, while we shall still have

$$4\pi\left(\dot{f}+\frac{4\pi\sigma}{K}f\right)=\frac{\partial}{\partial y}\left(\frac{c}{\mu}\right)-\frac{\partial}{\partial z}\left(\frac{b}{\mu}\right).$$

Putting $\sigma = 0$, and taking K and μ constant and $v_y = 0 = v_z$, we have

$$\frac{4\pi}{K}\left(\frac{\partial f}{\partial y}-\frac{\partial g}{\partial x}\right)=\dot{c}+v_x\frac{\partial c}{\partial x}$$

on the understanding that

$$c=\left(\frac{\partial G}{\partial x}-\frac{\partial F}{\partial y}\right);$$

similarly,
$$\frac{4\pi}{K}\left(\frac{\partial h}{\partial x}-\frac{\partial f}{\partial z}\right)=\dot{b}+v_x\frac{\partial b}{\partial x},$$

whence, if
$$\frac{\partial f}{\partial x}+\ldots+\ldots=0,$$

$$\frac{1}{K\mu}\nabla^2 f=\ddot{f}+v_x\frac{\partial \dot{f}}{\partial x}$$

Putting $\dfrac{1}{K\mu}=V^2$, where V is the velocity of light in the medium, we have

$$V^2\nabla^2 f=\ddot{f}+v_x\frac{\partial \dot{f}}{\partial x}.$$

214. The same result is at once obtained, if we admit that $\ddot{f}+\left(v_x\dfrac{\partial}{\partial x}+v_y\dfrac{\partial}{\partial y}+v_z\dfrac{\partial}{\partial z}\right)\dot{f}=$ the total differential of $\dot{f}$ with regard to t*. To solve this, when $v_y = 0 = v_z$, and a plane polarised light is travelling along x, put

$$f=\cos(qt-px),$$

whence
$$\frac{q}{p}=\frac{v_x}{2}\pm\sqrt{V^2+\frac{v_x^2}{4}},$$

or taking the upper sign corresponding to the propagation in the positive direction of X,

$$\frac{q}{p}=\frac{v_x}{2}+V,$$

approximately. This would be the velocity of propagation of light in the medium. The result, however, does not agree with experiment.

* J. J. Thomson, *Phil. Mag.* 1880.

215. On the other hand, on the elastic solid theory (labile-ether) assuming, as in Art. 102, an equation of the form

$$\rho\ddot{w} = n\nabla^2 w - \rho_1 \left(\frac{\partial}{\partial t} + v_x \frac{\partial}{\partial x} + v_y \frac{\partial}{\partial y} + v_z \frac{\partial}{\partial z}\right)^2 A w,$$

and putting $v_x = 0 = v_y$ (the direction of propagation being along z), we get

$$\rho\ddot{w} = n\nabla^2 w - \rho_1 A \left(\frac{\partial}{\partial t} + v_z \frac{\partial}{\partial z}\right)^2 w.$$

If we, now take $w = f(z - Vt)$, we get

$$\left(\frac{n}{\rho + \rho_1 A}\right)^{\frac{1}{2}} + \left(\frac{\rho_1 A}{\rho + \rho_1 A}\right) v_z = V, \text{ nearly}$$

i.e.
$$V = V_0 + v_z \frac{\lambda^2 - 1}{\lambda^2},$$

when λ is the refractive index, for

$$\lambda = \frac{\sqrt{\dfrac{n}{\rho}}}{\sqrt{\dfrac{n}{\rho + \rho_1 A}}}, \quad \text{i.e.} \quad \frac{\lambda^2 - 1}{\lambda^2} = \frac{\rho_1 A}{\rho + \rho_1 A}.$$

216. The same result follows from Fresnel s theory.

For this, let*

ρ = density of free ether.

ρ' = density of bound ether, i.e. ether dragged through free ether on account of motion of bodies through the ether.

V = velocity in free ether of propagation of light waves.

T = the periodic time.

V' = velocity in a medium consisting of free and bound ether at rest.

T' = the corresponding periodic time.

The elastic force (F) of restitution

$= \rho u \left(\dfrac{2\pi}{T}\right)^2$, as on Fresnel's theory of elastic ether

$= (\rho + \rho') u \left(\dfrac{2\pi}{T'}\right)^2$, assuming that the elastic force in both cases is the same, u being the displacement of the ether during the propagation of light

$$\therefore \quad (\rho + \rho') V'^2 = \rho V^2, \text{ i.e. } \rho + \rho = \rho\lambda^2.$$

* Potier, *Journal de Physique*, 1re série, t. v. (p. 105).

Suppose the condensed ether to move with the velocity v, in the direction of propagation of waves, V_1 = the velocity of propagation in free ether (when the medium of free and bound ether is in motion) of vibrations of the same amplitude and the same wave-length (l) as the preceding, and

$$V_1 - v = \text{velocity relative to condensed ether.}$$

Then $T_1 = \dfrac{l}{V_1}$, in free ether,

$$T_1' = \frac{l}{V_1 - v}, \text{ in the medium,}$$

and
$$F = \frac{4\pi^2}{T_1^2}\rho u + \frac{4\pi^2}{T_1'^2}\rho' u = \frac{4\pi^2}{l^2}u\left[\rho V_1^2 + \rho'(V_1 - v)^2\right].$$

$$\therefore \quad \rho V^2 = (\rho + \rho') V'^2 = \rho V_1^2 + \rho'(V_1 - v)^2.$$

If, now, ϵv = velocity of drift of luminous waves, $\epsilon v = V_1 - V'$,

i.e.
$$(\rho + \rho') V'^2 = \rho (V' + \epsilon v)^2 + \rho (V' + \epsilon v - v)^2,$$

$$\therefore \epsilon = \frac{\rho'}{\rho + \rho'} = \frac{(\lambda^2 - 1)}{\lambda^2},$$

as in Art. 215. This agrees with experiment. ϵ is called the dragging coefficient.

217. It is, *a priori*, evident therefore that a theory of moving electrons is likely to give a better account of the phenomena than the electro-magnetic theory in its simple form.

218. Before we proceed to consider this, however, it will be desirable to discuss certain modifications of the formulae proposed by Hertz, and see whether these will serve our purpose better.

Let us proceed, in the first place, to find

$$\frac{D}{Dt}\int (la + mb + nc)\, dS \equiv \frac{D}{Dt}\int R_n dS, \text{ say,}$$

when the medium of which S forms part is in motion or the total rate of variation $\left(\dfrac{D}{Dt}\right)$ in the number of tubes of induction. It will, of course, consist of two parts:

(1) $\displaystyle\int (l\dot{a} + m\dot{b} + n\dot{c})\, dS \equiv \int \frac{dR_n}{dt} dS$, say.

(2) Rate of change $\left(\dfrac{\partial}{\partial t}\right)$ depending on the velocity of the tubes.

Suppose the tubes across dS spread over dS' on account of this velocity; then the terms depending on the velocity

$$=\int (R_n' dS' - R_n dS).$$

Consider the volume enclosed by dS and dS'. The total surface integral over this surface is

$$-\left(\frac{\partial a}{\partial x}+\frac{\partial b}{\partial y}+\frac{\partial c}{\partial z}\right) dx\,dy\,dz.$$

But this must be *

$= - R_n' dS' + R_n dS +$ surface integral over the tubular surface generated by the contour of dS

Now, if $ds =$ an element of contour of dS (at x, y, z),

$U =$ velocity of a tube, so that

$Udt =$ length of the line joining corresponding points of dS and dS', and

$\epsilon =$ inclination of ds to the direction of the displacement of a tube;

we have $ds\, Udt \sin\epsilon =$ area of the portion of the tubular surface generated by ds.

Then, the surface integral over the tubular surface

$$=\int R_n ds\, dt\, U \sin\epsilon$$

$$=\int \begin{vmatrix} dx & dy & dz \\ a & b & c \\ v_x & v_y & v_z \end{vmatrix} dt,$$

where v_x, v_y, v_z are the component velocities of a tube corresponding to the velocity U,

$$\equiv \int (L dx + M dy + N dz)\, dt \text{ (say)},$$

so that $L, M, N = (bv_z - cv_y), (cv_x - av_z), (av_y - bv_x)$.

But

$$\int (L dx + M dy + N dz) = \int dS \left\{ l\left(\frac{\partial N}{\partial y} - \frac{\partial M}{\partial z}\right) + \ldots + \ldots \right\}.$$

Hence $\quad -\left(\frac{\partial a}{\partial x} + \ldots + \ldots\right) U\, dt\, dS \cos\theta$

$$= -\partial (R dS) + dS\, dt \left\{ l\left(\frac{\partial N}{\partial y} - \frac{\partial M}{\partial z}\right) + \ldots + \ldots \right\},$$

* Walker, *Outlines of the Theory of Electromagnetism.*

where θ = angle between the direction of motion of a tube and the normal to dS, and, accordingly,

$$U\cos\theta = lv_x + mv_y + nv_z.$$

Hence, we have

$$(lv_x + mv_y + nv_z)\left(\frac{\partial a}{\partial x} + \dots + \dots\right) dS\, dt$$

$$= \partial\,(RdS) - dS\,dt\left\{l\left(\frac{\partial N}{\partial y} - \frac{\partial M}{\partial z}\right) + \dots + \dots\right\}.$$

$$\therefore \quad \frac{D}{Dt}\int (la + mb + nc)\, dS$$

$$= \int (l\dot{a} + m\dot{b} + n\dot{c})\, dS + (lv_x + mv_y + nv_z)\left(\frac{\partial a}{\partial x} + \dots + \dots\right)$$

$$+ \int dS\left\{l\left(\frac{\partial N}{\partial y} - \frac{\partial M}{\partial z}\right) + \dots + \dots\right\} \quad \text{.........(1)}.$$

219. Thus, we get

$$\dot{a} + v_x\left(\frac{\partial a}{\partial x} + \dots + \dots\right) + \left(\frac{\partial N}{\partial y} - \frac{\partial M}{\partial z}\right)$$

$$= \frac{Da}{Dt}$$

$$= \left(\frac{\partial Y}{\partial z} - \frac{\partial Z}{\partial y}\right),\ \text{etc.} \quad \text{..............(2)},$$

since $\quad -\dfrac{D}{Dt}\displaystyle\int (la + mb + nc)\, dS = \int (Xdx + Ydy + Zdz).$

The equations given by Hertz are

$$\frac{d\mu a}{dt} + v_x\left(\frac{\partial \mu a}{\partial x} + \dots + \dots\right) + \left(\frac{\partial N}{\partial y} - \frac{\partial M}{\partial z}\right) = \frac{D(\mu a)}{Dt},\ \text{etc....(3)}.$$

Similarly,

$$\frac{D}{Dt}(KX) = \frac{\partial}{\partial t}(KX) + v_x\left(\frac{\partial (KX)}{\partial x} + \dots + \dots\right) + \left(\frac{\partial N'}{\partial y} - \frac{\partial M'}{\partial z}\right),$$

etc. ....(4)

where $\quad L' = K(v_z Y - v_y Z)$, etc.

220. Again, since

$$4\pi i = \int(\alpha dx + \beta dy + \gamma dz)$$

$$= 4\pi \int[(p + \dot{f})\, l + \dots + \dots]\, dS,$$

we have $\quad -\dfrac{\partial \beta}{\partial z} + \dfrac{\partial \gamma}{\partial y} = 4\pi\,(p + \dot{f})$, etc.

When the medium is in motion $\dot{f}$ should be changed into $\frac{Df}{Dt}$, $\therefore$ the total current will be $p + \frac{Df}{Dt}$

$$= p + \dot{f} + v_x \left(\frac{\partial f}{\partial x} + \dots + \dots\right) + \left(\frac{\partial N''}{\partial y} - \frac{\partial M''}{\partial z}\right), \text{ etc.},$$

where $$L'' \equiv (gv_z - v_y h), \text{ etc.}$$

$$= p + \dot{f} + v_x\rho + \frac{\partial N''}{\partial y} - \frac{\partial M''}{\partial z}, \text{ etc.}$$

Thus the total current will be the sum of

(1) Conduction current (p),

(2) Displacement current ($\dot{f}$),

(3) Convection current ($v_x\rho$) [Art. 227],

(4) Current which may be identified with that discovered by Röntgen*.

221. Let us proceed now to apply the above equations to aberration.

Let us suppose the wave to be propagated along x with velocity v_x, so that $\alpha = 0$, $X = 0$, and let us further suppose $\gamma = 0$, $Y = 0$ (since the magnetic and electric forces should be at right angles to each other)

Then, putting $$\mu = 1,$$

$$\beta = \psi(x - V't),$$

and $$Z = \phi(x - V't),$$

since $$\dot{\beta} = -V'\psi' \quad \dot{Z} = -V'\phi'$$

$$\frac{\partial \beta}{\partial x} = \psi', \quad \frac{\partial Z}{\partial x} = \phi'$$

Substituting in the equation (2) [Art. 219], we get

$$\dot{\beta} + v_y\left(\frac{\partial \alpha}{\partial x} + \dots + \dots\right) + \frac{\partial}{\partial z}(\beta v_z) + \frac{\partial}{\partial x}(\beta v_x) = \frac{\partial Z}{\partial x}, \text{ etc.},$$

where (v_x, v_y, v_z) is the velocity of the medium

If $v_y = 0 = v_z$, we have

$$-V'\psi' + v_x\psi' = \phi'$$

* *Phil. Mag.* May, 1888.

Therefore, from the equation

$$\frac{Dh}{Dt} = \frac{D}{Dt}(KZ) = \frac{\partial}{\partial t}(KZ) + v_z\left(\frac{\partial\,(KZ)}{\partial x} + \ldots + \ldots\right) + \ldots$$

where $$\frac{Dh}{Dt} = \left(-\frac{\partial \alpha}{\partial y} + \frac{\partial \beta}{\partial x}\right) - 4\pi r, \text{ from Art. 220,}$$

we get $$K\left(\dot{Z} + v_x\frac{\partial Z}{\partial x}\right) = \frac{\partial \beta}{\partial x},$$

or $$V' = \frac{1}{\sqrt{K}} + v_x.$$

Thus, the modified equation still fails to give the correct result*.

* Poincaré, *Électricité et Optique.*

CHAPTER IV

ELECTRON THEORY

SECTION I. THEORETICAL

222. From the investigations of the last chapter, we conclude that the electro-magnetic theory, as expressed by the equations of Maxwell and Hertz, cannot account for aberration, dispersion and allied phenomena. In analysing the reason for this, we note that the theory is based on the following postulates:

(1) The energy of the electro-magnetic field is that of the dielectric medium alone, arising from a certain strained condition of the medium.

(2) The conductors having static charges serve only to limit the dielectric region, so that no part of the energy resides on it.

(3) The strained condition of a dielectric is due to electric displacement or polarisation (f, g, h), subject to the condition

$$\frac{\partial f}{\partial x}+\frac{\partial g}{\partial y}+\frac{\partial h}{\partial z}=0 \quad \ldots\ldots\ldots\ldots\ldots\ldots\ldots(1).$$

This displacement is apparently held to involve motion of the ether in the medium, subject to a property akin to elasticity (due to interaction of matter and ether) defined by the so-called specific inductive capacity of the medium which thus appears as a constant of the medium.

(4) Conduction as well as convection currents (as in the case of an electrical discharge) involve a transference ("a procession and not an arrangement" as Faraday put it) dependent on a certain property of the conductor (called its

conductivity). This transference is that of electric charge, but what this charge is—whether it is material or ethereal—is not further specified or is rather left entirely open*.

223. If we limit ourselves to two physical entities, matter and ether, the electric charge whose motion constitutes electric current must be regarded as a mode of manifestation of the ether. If, on the other hand, we agree to regard a unit of electric charge as an actual physical entity, distinct from matter and ether (but related to them and partaking the nature of both in a manner that will require further investigation), we are able to give an account of the various phenomena which are left unexplained on the above postulates.

224. We are led to this additional postulate of a unit of electric charge as a physical entity not merely on theoretical grounds, but as a result of direct experiment. For during electrolysis, each monovalent atom is known to carry with it to the anode a determined quantity of what has hitherto been called a negative electric charge which can be measured and which is independent of the nature of the transported atom. If we assume that this also is the unit charge, which takes part in electric conduction or convection, and if we call it an 'electron,' we must conceive a monovalent material atom showing no electrical properties as the result of combination of a single electron with what may fittingly be called one or more units of 'positive electricity.' This would amount then to the statement that the electrical properties of bodies (as well as those of a dielectric medium) are due to the presence of 'electrons' associated with atoms of matter, forming systems of various complexity, and such a theory is found to be consistent with observed facts.

* Maxwell's pronouncement on this point, indeed, clearly sets forth the position: *Electricity and Magnetism*, Vol. I.

"It appears to me that while we derive great advantage from the recognition of the many analogies between electric current and current of material fluid, we must carefully avoid making any assumption, not warranted by experimental evidence, and that there is as yet no experimental evidence to show whether the electric current is really a current of a material substance or a double current, or whether its velocity is great or small as measured in feet per second."

225. Maxwell's remarks [Art. 222 (4) note] had reference to the negative results of the experiments whereby he proposed to detect in a direct manner the inertia of an electric charge. Experimental determination of the mass of an electron as well as its velocity under certain conditions* has supplied the data sought by Maxwell on which to build up the further development of the electro-magnetic theory.

226. We have thus justification for regarding the phenomenon of electric conduction in gases as due to 'ionisation' or generation of charged particles or 'ions' which are carriers of electricity.

227. An electric current would then, on this view, consist of two parts—one due to electric displacement which we may still regard as ethereal and the other due to motion of electrons; or if u is the total current, in the direction of x, we shall have (ignoring Röntgen current)

$$u = \dot{f} + \rho\dot{x} \qquad (2),$$

where ρ is the volume density of (free) electrification whose transfer or procession gives rise to convection or conduction currents, and $\dot{x}$ its velocity in the direction of x, while

$$\dot{f} \equiv \frac{df}{dt}.$$

228. Now, when there is free electrification, we must have

$$\frac{\partial f}{\partial x} + \frac{\partial g}{\partial y} + \frac{\partial h}{\partial z} = \rho \qquad (3).$$

This must therefore be the further condition satisfied by f, g, h, while in the case of conduction there must, in addition, be a viscous decay which has to be taken account of in a suitable dissipation function.

229. When we proceed to interpret these equations, coupled with the observed phenomena of metallic conduction and dielectric polarisation, in terms of the electron theory, we are naturally led to the conclusion that:

230. A dielectric medium must be conceived to have electrons interspersed in it, giving rise to a constrained (ethereal) motion in the medium defined by the above relation.

* J. J. Thomson, *Conduction of Electricity through Gases*, Ch. v.

This also necessarily imposes a constraint on the motion of electrons. On the other hand, motion of electrons in a conductor must be free (subject however to certain dissipation of energy), so that the static charge in them always resides on the surface.

231. A conception of this kind naturally suggests that the dielectric property of a medium may be explained as arising from the constrained motion imposed on it. We proceed to show that this is possible and that in this way we can explain dispersion and allied phenomena.

232. On comparing the equations (1) and (3), we readily see that these cannot be, obviously, satisfied at the same time if f, g, h are to have the same meaning in both.

233. Now if $\rho = 0$, that is in *free ether*,

$$\frac{\partial f}{\partial x} + \frac{\partial g}{\partial y} + \frac{\partial h}{\partial z} = 0.$$

Thus, we may say that in free ether

$$\frac{\partial f}{\partial x} + \frac{\partial g}{\partial y} + \frac{\partial h}{\partial z} = 0,$$

that in a material medium

$$\frac{\partial f}{\partial x} + \frac{\partial g}{\partial y} + \frac{\partial h}{\partial z} = \rho,$$

while in all cases

$$\frac{\partial f_0}{\partial x} + \frac{\partial g_0}{\partial y} + \frac{\partial h_0}{\partial z} = 0 \quad \text{.....................(4)},$$

where f_0, g_0, h_0 are defined by the equation (4) and are equal to f, g, h, when the medium is free ether.

234. In order to specify these quantities further, we observe that from the equations

$$\frac{\partial f_0}{\partial x} + \frac{\partial g_0}{\partial y} + \frac{\partial h_0}{\partial z} = 0 \quad \text{.....................(4)},$$

and

$$\frac{\partial f}{\partial x} + \frac{\partial g}{\partial y} + \frac{\partial h}{\partial z} = \rho \quad \text{.....................(3)},$$

we may write

$$f_0 = f + A, \text{ etc.} \quad \text{.....................(5)},$$

where

$$\frac{\partial A}{\partial x} + \frac{\partial B}{\partial y} + \frac{\partial C}{\partial z} = -\rho \quad \text{.....................(6)}.$$

This expression suggests that A, B, C are components of a quantity I which corresponds to the coefficient of magnetisation on the usual theory of magnetism, so that, just as dM is magnetic moment of an elementary magnet $= Id\tau$ (where $d\tau$ is an element of volume), similarly, we may take $dM =$ electric moment of an element of volume containing electric charges arising from the presence of electrons and

$$dM = Id\tau,$$

where of course $$I = (A, B, C) \quad\text{.......................(7).}$$

235. Again, from (5), we get

$$\dot{f}_0 = \dot{f} + \dot{A} \quad\text{..........................(8).}$$

And if we agree that $\dot{f}_0 = u =$ total current (polarisation and convection), we conclude that

$$\dot{A} = \rho\dot{x} \quad\text{.........................(9),}$$

$(\dot{x}, \dot{y}, \dot{z})$ being the velocity of electrons, as before, and it is easily seen that (7) and (9) are consistent, on the understanding that

$$\rho(x - x_0) = A \text{ (say)} \quad\text{or}\quad e(x - x_0) = Ad\tau,$$

e being the invariable charge in volume $d\tau$, and $x - x_0 =$ the displacement of e in the direction of x.

Comparing the equations (3), (4), (5) and (6), we conclude that total polarisation in any medium may be regarded as made up of two parts—one involving ethereal (f, g, h) and the other corpuscular (A, B, C) displacement.

Again, comparing (8) and (3), we observe that the total current is to be regarded as similarly made up and that it is the total current and total polarisation that are subject to the solenoidal condition.

From (6) since, in stationary media,

$$\frac{\partial \dot{A}}{\partial x} + \frac{\partial \dot{B}}{\partial y} + \frac{\partial \dot{C}}{\partial z} + \frac{\partial \rho}{\partial t} = 0,$$

we get the equation of continuity for electrons in motion, viz.

$$\frac{\partial \rho}{\partial t} + \frac{\partial \rho\dot{x}}{\partial x} + \ldots + \ldots = 0.$$

Again, from (6), we can obviously derive

$$A = \lambda \frac{\partial \phi}{\partial x}, \text{ etc.}$$

where λ is defined by $\lambda \nabla^2 \phi + \rho = 0$.

The equation states that ϕ is the potential of a distribution ρ, provided

$$\lambda = \frac{1}{4\pi} k_0,$$

where k_0 is the specific inductive capacity of the medium. Since the medium here is the ethereal medium (whose property is modified by the presence of electrons), k_0 is an absolute constant defining the property of the free ethereal medium.

Thus

$$A = \frac{k_0}{4\pi} \cdot \frac{\partial \phi}{\partial x}, \text{ etc., and } f_0 = f + \frac{k_0}{4\pi} \cdot \frac{\partial \phi}{\partial x}, \text{ etc.} \quad \ldots(10),$$

so that

$$f_0 = f + k_0 \frac{e}{4\pi} \cdot \frac{\partial}{\partial x} \left(\frac{1}{r}\right)$$

in the neighbourhood of an electric charge e.

236. Again, on analysing the electro-magnetic theory of magnetism, we find that:

(1) Maxwell takes magnetic force, though invariably associated with and definitely related to electricity in motion, as a separate physical entity;

(2) and distinguishes between magnetic force and magnetic induction by means of a quantity μ which defines the magnetic property of the medium and is taken to be practically constant, though not specifically so. But inasmuch as, on experimental grounds, μ, as thus defined, depends on the magnetic force itself, one can hardly regard such a course justifiable. Hertz takes it as a variable in his equations, but one finds it difficult to follow the consequences of such a hypothesis.

It will be seen, however, that an immediate simplification results, as soon as the intimate relation between electricity and magnetism is recognized, and an attempt made to describe the effect of one in terms of the other. The first attempt in this direction was eminently fruitful in giving a more coherent view of the electro-magnetic field than would otherwise have been possible. This was to regard an elementary magnet as identical with a molecular electric current and a magnetic shell, the same as an electric circuit coinciding with its periphery, being distinguished from the latter only in this that while the first is a simply-connected region, the second is doubly-connected.

The next step in simplification would obviously be not to consider magnetism as a separate entity at all, but to regard it as a condition of the medium, in which electric action is taking place—defined by the equations of Maxwell, involving magnetic force. Since an electric current gives rise to a magnetic field, magnetism would, on this view, correspond to a motion of the ethereal medium, due to movement associated with an electric current.

Finally, the optical medium being admitted to be the same as the electro-magnetic field, optical phenomena must be held to be associated with the (ethereal) motion which constitutes magnetism.

237. On this understanding, the so-called electrostatic energy of the ether will be W_1, where

$$W_1 = \frac{2\pi}{k_0}\int (f^2 + g^2 + h^2)\, d\tau \ldots\ldots\ldots\ldots\ldots(11),$$

while the potential energy of the material medium will be W, where

$$W = \frac{2\pi}{K}\int (f_0^2 + g_0^2 + h_0^2)\, d\tau \ldots\ldots\ldots\ldots\ldots(12),$$

while the kinetic energy (T) of a material medium is to be taken equal to

$$\frac{\mu}{8\pi}\int (\alpha^2 + \beta^2 + \gamma^2)\, d\tau \ldots\ldots\ldots\ldots\ldots\ldots(13).$$

238. From (10) we observe, what is *a priori* evident, that the total force producing the strained condition of the medium is made up of the force associated with the free ethereal motion and that due to electronic disturbance.

From the expressions for W and T, (12) and (13), applying $\delta\int (T - W)\, d\tau = 0$, where δ is the operator of the calculus of variation, and remembering that

$$4\pi \dot{f}_0 = \left(\frac{\partial \gamma}{\partial y} - \frac{\partial \beta}{\partial z}\right) \ldots\ldots\ldots\ldots\ldots\ldots\ldots(14),$$

we get $$\ddot{f_0} = V^2\nabla^2 f_0, \quad \left[K\mu = \frac{1}{V^2}\right] \quad \text{.............. (15).}$$

But $$\dot{f_0} = \dot{f} + \dot{A} = \dot{f} + \rho\dot{x} \quad \text{....................(16),}$$

$$\therefore \quad \nabla^2 f_0 = \nabla^2 f + \nabla^2 A = \nabla^2 f + \frac{k_0}{4\pi}\frac{\partial}{\partial x}\nabla^2\phi \quad \text{......(17),}$$

or $$\ddot{f} + \frac{d}{dt}(\rho\dot{x}) = V^2\left(\nabla^2 f - \frac{\partial\rho}{\partial x}\right) \quad \text{...........(18).}$$

Also from Art. 163, we get

$$-\frac{d}{dt}\int (la + mb + nc)\, dS = \int (X\,dx + Y dy + Z dz) \quad \text{...(19),}$$

where (a, b, c) = magnetic induction,

(X, Y, Z) = electric force.

Hence we have, if

$$X = \frac{4\pi}{k_0} f \quad [\text{Art } 235],$$

$$\dot{\alpha} = -4\pi V_0^2\left(\frac{\partial h}{\partial y} - \frac{\partial g}{\partial z}\right), \text{ where } k_0 = \frac{1}{V_0^2} \quad \text{........ (20)}$$

(on the understanding that $\mu = 1$), so that V_0 is the velocity of propagation in free ether.

From (14), (16) and (20), we get

$$\ddot{\alpha} = -4\pi V_0^2\left(\frac{\partial \dot{h}}{\partial y} - \frac{\partial \dot{g}}{\partial z}\right) \quad \text{.................(21)}$$

in stationary media

$$= V_0^2\left[\nabla^2\alpha - \frac{\partial}{\partial x}\left(\frac{\partial\alpha}{\partial x} + \frac{\partial\beta}{\partial y} + \frac{\partial\gamma}{\partial z}\right) + \frac{\partial\rho\dot{z}}{\partial y} - \frac{\partial\rho\dot{y}}{\partial z}\right] \quad \text{...(22).}$$

The equations (18) and (22) are the fundamental equations of Lorentz's theory which, it should be noted, were otherwise obtained in Ch. III.

239. Let, now, T_1 = the kinetic energy of an electron (e),
T_2 = the kinetic energy of the ether,
V_2 = the potential energy of the ether,
V_1 = the potential energy of extraneous forces.

Then, since $-\int(\beta h - \gamma g)\,d\tau =$ momentum of the field ($\mu = 1$) in the direction of x [J. J. Thomson, *Recent Researches*],

$$-\frac{d}{dt}\int(\beta h - \gamma g)\,d\tau = \frac{d}{dt}\left(\frac{\partial T_2}{\partial \dot{x}}\right) \text{ [Art. 243]},$$

x, y, z defining the position of an electron.

Applying Lagrange's equations of motion for the system consisting of the ether and the electron, viz.

$$\frac{d}{dt}\left(\frac{\partial T_1}{\partial \dot{x}} + \frac{\partial T_2}{\partial \dot{x}}\right) - \frac{\partial (T_1 + T_2)}{\partial x} + \frac{\partial (V_1 + V_2)}{\partial x} = 0,$$

we get $$\frac{d}{dt}\left(\frac{\partial T_1}{\partial \dot{x}}\right) + \frac{\partial V_1}{\partial x} = \frac{\partial T_2}{\partial x} + \frac{d}{dt}\int(\beta h - \gamma g)\,d\tau - \frac{\partial V_2}{\partial x},$$

since $$\frac{\partial T_1}{\partial x} = 0, \text{ obviously;}$$

i.e. $$\frac{d}{dt}\left(\frac{\partial T_1}{\partial \dot{x}}\right) + \frac{\partial V_1}{\partial x} = \int(h\dot{\beta} - g\dot{\gamma})\,d\tau + \int(\beta\dot{h} - \gamma\dot{g})\,d\tau + \frac{\partial T_2}{\partial x} - \frac{\partial V_2}{\partial x}$$

$$= -\frac{4\pi}{k_0}\int\left[\left(\frac{\partial f}{\partial z} - \frac{\partial h}{\partial x}\right)h - g\left(\frac{\partial g}{\partial x} - \frac{\partial f}{\partial y}\right)\right]d\tau$$

$$- \int[\beta(w - \rho\dot{z}) - \gamma(v - \rho\dot{y})]\,d\tau + \frac{\partial T_2}{\partial x} - \frac{\partial V_2}{\partial x} \quad \text{......(23)},$$

since $$-\dot{\beta} = \frac{4\pi}{k_0}\left(\frac{\partial f}{\partial z} - \frac{\partial h}{\partial x}\right) \text{ and } w = \dot{h} + \rho\dot{z}, \text{ etc. [Art. 227].}$$

Integrating by parts the first integral of (23) and substituting

$$4\pi w = \frac{\partial \beta}{\partial z} - \frac{\partial \gamma}{\partial y}, \text{ etc.,}$$

in the second integral and again integrating by parts, we get

$$\frac{d}{dt}\left(\frac{\partial T_1}{\partial \dot{x}}\right) + \frac{\partial V_1}{\partial x} = \frac{4\pi}{k_0}\int f\left(\frac{\partial f}{\partial x} + \frac{\partial g}{\partial y} + \frac{\partial h}{\partial z}\right)d\tau - \frac{\partial T_2}{\partial x} + \frac{\partial T_2}{\partial x}$$

$$+ \int \rho(\dot{y}\gamma - \dot{z}\beta)\,d\tau + \frac{\partial V_2}{\partial x} - \frac{\partial V_2}{\partial x}$$

$$= \frac{4\pi}{k_0}\int \rho f\,d\tau + \int \rho(\dot{y}\gamma - \dot{z}\beta)\,d\tau \quad \text{.........(24)},$$

which may be stated in words as follows:—

An electron in motion is subjected to a force which is the sum of an electrostatic force and an electrodynamic force, as the result of the action of the field*.

* Poincaré, *Électricité et Optique*.

240. Again, T_2 being the kinetic energy of free ether, we have

$$T_2 = \int (Fu + Gv + Hw)\, d\tau,$$

where $u = \dot{f} + \dot{A}$, etc., while

$$W = \text{the potential energy} = \frac{2\pi}{k_0} \int (f^2 + g^2 + h^2)\, d\tau,$$

where (it should be remembered) $k_0 = \frac{1}{V_0^2}$ (V_0 being the velocity of light in free ether) and f, g, h are connected by the equation of condition

$$\frac{\partial f}{\partial x} + \frac{\partial g}{\partial y} + \frac{\partial h}{\partial z} = \rho.$$

Introducing (in the manner of Larmor) an undetermined multiplier ϕ and writing

$$W' = \frac{2\pi}{k_0} \int (f^2 + g^2 + h^2) d\tau - \int \phi \left[\left(\frac{\partial f}{\partial x} + \frac{\partial g}{\partial y} + \frac{\partial h}{\partial z}\right) - \rho\right] d\tau,$$

we have

$$\delta W' = \frac{4\pi}{k_0} \int (f\delta f + \ldots + \ldots)\, d\tau - \int \phi \left[\left(\frac{\partial \delta f}{\partial x} + \ldots + \ldots\right) - \delta\rho\right] d\tau \qquad \text{.........(25)},$$

since ϕ is not to vary. Now integrating by parts into surface (S) integrals and volume integrals, we get

$$\delta W' = \int \left\{\left(\frac{4\pi}{k_0} f + \frac{\partial \phi}{\partial x}\right) \delta f + \ldots + \ldots\right\} d\tau + \int \phi\, \delta\rho\, d\tau - \int \phi\, (lf + \ldots + \ldots)\, dS.$$

Hence, from Lagrange's equation (since f, g, h may be regarded as generalized coordinates) we have

$$\frac{d}{dt} \frac{\partial T_2}{\partial \dot{f}} - \frac{\partial T_2}{\partial f} + \frac{\partial W'}{\partial f} = 0$$

or

$$\dot{F} + \frac{4\pi}{k_0} f + \frac{\partial \phi}{\partial x} = 0 \qquad \text{..................(26)},$$

where ϕ is evidently the potential of a free distribution ρ.

That is (if the medium is at rest), the electrostatic force due to the medium (say X)

$$= -\dot{F} - \frac{\partial\phi}{\partial x}$$

(the latter of which is due to ρ) on the understanding that

$$\frac{4\pi}{k_0} f = X.$$

241. It will be observed that K of the ordinary electromagnetic theory becomes k_0. This amounts to the postulate that there is no dielectric other than free ether [Art. 230].

Remembering that $A = \frac{k_0}{4\pi} \frac{\partial\phi}{\partial x}$,

we conclude that $\dot{F} = -\frac{4\pi}{k_0} f_0$

or F is the total electro-kinetic momentum of the medium.

242. We may in fact picture to ourselves the intimate processes going on in a medium, in which a transfer of electrical or optical energy is taking place by means of the moving Faraday tubes of J. J. Thomson. On this view, the Faraday tubes are supposed to move with the velocity of light and both electric and magnetic forces and electric polarisation and therefore the state of energy of the field are completely represented by these motions*.

243. To show this, let the electric displacement across any surface be equal to the difference in the number of tubes that enter and leave the surface, and f, g, h their components parallel to the axes, and l, m, n the direction-cosines of an element of surface.

Then, since in any limited region, the total number of tubes must remain unchanged,

$$\frac{D}{Dt}\int (lf + mg + nh)\, dS = 0 \equiv \frac{D}{Dt}\int R_n dS \text{ (say)}$$

$$= \int (l\dot{f} + m\dot{g} + n\dot{h})\, dS + \text{terms depending on the velocity of the moving tubes}$$

$$\left[\dot{f} \equiv \frac{df}{dt}, \text{ etc.}\right].$$

* *Recent Researches*, Ch. I.

Then exactly as in Art. 218, writing f, g, h for a, b, c, and

$$L, M, N = (gv_z - v_y h),\quad (hv_x - v_z f),\quad (fv_y - v_x g),$$

we have

$$(lv_x + mv_y + nv_z)\left(\frac{\partial f}{\partial x} + \dots + \dots\right) dS\,.\,dt$$

$$= \partial (R_n dS) - dS\,.\,dt \left\{ l\left(\frac{\partial N}{\partial y} - \frac{\partial M}{\partial z}\right) + \dots + \dots \right\};$$

$$\therefore\quad \frac{D}{Dt}\int (lf + mg + nh)\, dS$$

$$= \int (l\dot{f} + m\dot{g} + n\dot{h})\, dS + (lv_x + mv_y + nv_z)\left(\frac{\partial f}{\partial x} + \dots + \dots\right)$$

$$+ \int dS \left\{ l\left(\frac{\partial N}{\partial y} - \frac{\partial M}{\partial z}\right) + \dots + \dots \right\} = 0.$$

We get, accordingly,

$$\frac{df}{dt} = \dot{f} = -v_x\left(\frac{\partial f}{\partial x} + \frac{\partial g}{\partial y} + \frac{\partial h}{\partial z}\right) + \frac{\partial}{\partial y}(v_x g - v_y f) - \frac{\partial}{\partial z}(v_z f - v_y h).$$

But $\frac{\partial f}{\partial x} + \frac{\partial g}{\partial y} + \frac{\partial h}{\partial z} = \rho =$ volume density of electrification;

$$\therefore\quad \dot{f} + v_x\rho = \frac{\partial}{\partial y}(v_x g - v_y f) - \frac{\partial}{\partial z}(v_z f - v_y h)$$

$$= u \text{ [Art. 227]},$$

where u is the total current in the direction of x.

But since, on Maxwell's theory,

$$4\pi u = \frac{\partial \gamma}{\partial y} - \frac{\partial \beta}{\partial z},\ \text{etc.},$$

where α, β, γ are the components of magnetic force, we may take as a particular solution,

$$\alpha = 4\pi (v_y h - v_z g),\ \text{etc.},$$

showing that Faraday tubes in motion give rise to a magnetic field.

Moreover, since the kinetic energy T_2 (per unit volume)

$$= \frac{1}{8\pi}(\alpha^2 + \beta^2 + \gamma^2),$$

we have the momentum of the tubes, or the momentum of the medium in the direction of x,

$$= \frac{\partial T_2}{\partial v_x} = \frac{1}{4\pi}\left(\alpha\frac{\partial \alpha}{\partial v_x} + \dots + \dots\right)$$

$$= (g\gamma - h\beta).$$

244. Thus, the density of the ether being assumed to be constant, $(\beta h - \gamma g)$ must be proportional to the velocity (say in the direction of x) of the ether. If the field is doubled, the velocity has to be quadrupled—a conclusion which does not seem to be altogether satisfactory.

In any case an actual transference of momentum from ether to matter is a fact of experience. For we know that a pressure is exerted by electric waves, including light waves, on a slab of a substance which absorbs energy.

The energy of radiation $= \frac{1}{8\pi\mu} B_0^2$, if $B_0 =$ the mean magnetic induction.

Therefore the momentum absorbed per unit of time

$$= \frac{1}{8\pi\mu} B_0^2.$$

For total reflection, the pressure $= \frac{1}{4\pi\mu} B_0^2$. This has been detected and measured by Lebedew and Nichols and Hall.

245. In order to prove that the total energy-density is equal to the pressure of radiation, we notice that the displacements corresponding to incident and reflected waves being ξ and ξ', we have

$$\xi = a \cos \frac{2\pi}{\lambda} (ct + x),$$

$$\xi' = a' \cos \frac{2\pi}{\lambda'} (ct - x),$$

x being in the direction of propagation.

Then, if we suppose the reflecting surface to move with velocity v, the condition of continuity gives, at $x = vt$,

$$\xi + \xi' = 0,$$

whence $$a \cos \frac{2\pi}{\lambda} (ct + vt) + a' \cos \frac{2\pi}{\lambda'} (ct - vt) = 0.$$

This requires $$a + a' = 0,$$

$$\frac{c+v}{\lambda} = \frac{c-v}{\lambda'}.$$

The energy of the incident wave per unit area corresponding to a length λ

$$=\frac{1}{2}\int_0^\lambda \rho\dot{\xi}^2 dx = \rho\frac{a^2c^2 . \pi^2}{\lambda^2}\int_0^\lambda\left[1-\cos\frac{4\pi}{\lambda}(ct+x)\right]dx$$

$$=\frac{\rho a^2 . \pi^2 c^2 \lambda}{\lambda^2}$$

$$\equiv t_i \text{ (say)};$$

$$\therefore\quad \frac{t_i}{t_r}=\frac{\lambda'}{\lambda}=\frac{c-v}{c+v}.$$

If p is the pressure of radiation,

$$t_r - t_i = pv = t_i\left(\frac{c+v}{c-v}-1\right),$$

where t_r is the energy over a length $c+v$, which is the displacement in unit time.

$$\therefore\quad t_i\frac{2}{c-v}=p.$$

If δ_1 = density of energy due to incident wave,

$$\delta_1=\frac{t_i}{c+v}.$$

Similarly
$$\delta_2=\frac{t_r}{c-v}=t_i . \frac{c+v}{(c-v)^2};$$

$$\therefore\quad \delta = \text{total energy-density} = \delta_1+\delta_2$$

$$=t_i\left(\frac{1}{c+v}+\frac{c+v}{(c-v)^2}\right)$$

$$=t_i\frac{(c-v)^2+(c+v)^2}{(c+v)(c-v)^2}=\frac{2t_i}{c-v}.\frac{c^2+v^2}{c^2-v^2}$$

$$=p.\frac{c^2+v^2}{c^2-v^2}$$

$$=p,$$

when v is small compared with c*.

246. The system of momenta of Art. 243 is found to be "proportional to amounts of energy transferred in unit time across unit planes at right angles to the axes of x, y, z on Poynting's theory of the transfer of energy in the electromagnetic field." Hence the direction of momentum† coincides with the direction of flow of energy on this theory.

* Larmor, *Encyc. Brit.* 'Radiation.'

† Note that we get the same system of momenta on the electron theory.

Again, since

$$T=\frac{\mu}{8\pi}(\alpha^2+\beta^2+\gamma^2)=2\pi\mu\{(hv_y-gv_z)^2+\ldots+\ldots\},$$

we have $$X=v_zb-v_yc=\frac{4\pi f}{K}, \text{ etc.},$$

if electromotive intensity (X, Y, Z) is due to the motion of Faraday tubes only, so that $\frac{\partial T}{\partial f}=X$, etc.

And since $$b=4\pi\mu(fv_z-hv_x),$$
$$c=4\pi\mu(gv_x-fv_y),$$

we have

$$f=\mu K\{f(v_x^2+v_y^2+v_z^2)-v_x(fv_x+gv_y+hv_z)\}, \text{ etc.}$$

Therefore multiplying in order by v_x, v_y, v_z, we have

$$fv_x+gv_y+hv_z=0,$$

and therefore $$v_x^2+v_y^2+v_z^2=\frac{1}{\mu K}=V^2,$$

where V is the velocity of light in the medium.

247. That is, the tubes move at right angles to the direction of polarisation and also at right angles to the direction of the magnetic force with the velocity of light. The energy of the electro-magnetic field is seen, therefore, on this view to be entirely kinetic.

248. We have seen that electrons are not hypothetical entities. The phenomena of electrolysis furnish evidence of the atomic nature of electricity, while the phenomena of ionisation prove the existence of electrified units. We have, therefore, to take account of them, apart from any simplification that their introduction brings about in our theoretical discussion: this simplification is, however, immediately noticeable, for, on this theory, magnetism is explained simply as the condition of the medium, due to the motion of electrons [Art. 236].

249. On Lorentz's theory, (*a*) there is no magnetism, (*b*) Ampèrean current is current due to motion of electrons, (*c*) in a conductor there is free motion of electrons, (*d*) in a dielectric, electrons cannot move far away from the position of equilibrium.

On the electron theory, if u, v, w = the total current,

$$u = \dot{f} + \rho\xi,$$

where $\rho\xi$ = current of convection, ξ, η, ζ being the velocity of an electron, and

$$\frac{\partial f}{\partial x} + \frac{\partial g}{\partial y} + \frac{\partial h}{\partial z} = \rho.$$

Then u, v, w will satisfy the relation

$$\frac{\partial u}{\partial x} + \frac{\partial v}{\partial y} + \frac{\partial w}{\partial z} = 0 \quad \text{.....................(27)},$$

for the left-hand expression is the same as

$$\frac{\partial}{\partial t}\left(\frac{\partial f}{\partial x} + \frac{\partial g}{\partial y} + \dots\right) + \left(\xi\frac{\partial \rho}{\partial x} + \dots + \dots\right) + \rho\left(\frac{\partial \xi}{\partial x} + \dots + \dots\right)$$

or
$$\frac{\partial \rho}{\partial t} + \xi\frac{\partial \rho}{\partial x} + \dots + \dots \equiv \frac{D\rho}{Dt},$$

since
$$\frac{\partial \xi}{\partial x} + \dots + \dots = 0,$$

if the dilatation of the particles is zero; but we may take

$$\frac{D\rho}{Dt} = 0,$$

if we suppose electrons to undergo no change in time, in respect of their charges. Thus, the relation (27) is satisfied.

Again, $\quad \nabla^2 F = -4\pi u$ [Art. 161],

since μ is to be taken equal to unity;

$$\therefore \quad \nabla^2\left(\frac{\partial F}{\partial x} + \frac{\partial G}{\partial y} + \frac{\partial H}{\partial z}\right) = -4\pi\left(\frac{\partial u}{\partial x} + \frac{\partial v}{\partial y} + \frac{\partial w}{\partial z}\right),$$

or
$$\frac{\partial F}{\partial x} + \frac{\partial G}{\partial y} + \frac{\partial H}{\partial z} = \int \frac{1}{r}\left(\frac{\partial u}{\partial x} + \frac{\partial v}{\partial y} + \frac{\partial w}{\partial z}\right) d\tau = 0.$$

This is the further relation to which the electron theory necessarily leads.

250. A suitable hypothesis in terms of moving electrons may be made to fit in with the Ampèrean theory of magnetism. For this, we need only conceive magnetic molecules as consisting of doublets of electrons circulating rapidly in the Ampèrean channels, and in order to account for the neutralization of their attraction we must conceive them revolving*.

If e = ionic charge, n = number of ions per unit length,

* Larmor, *loc. cit.* See, however, a paper by Sir J. J. Thomson, *Phil. Mag.* Oct. 1920.

A = the area embraced by the channels, I = intensity of magnetisation, l = length of the orbit, q = velocity of an ion, then

$$IAl = \text{magnetic moment} = \text{current} \times A = qnAe \times A\,;$$

$$\therefore\quad I = \frac{qne}{l}A.$$

But from electro-chemical data $en = 10^3$, and from molecular dimensions $\dfrac{A}{l} = \frac{1}{2}10^{-8}$.

Thus an intensity of magnetisation equal to 1700 C.G.S. units will mean a velocity of about 3×10^8, that is about $\frac{1}{100}$ of the velocity of light. Measurement of the velocities of cathode rays yields the same order of quantities*.

251. A magnetic molecule therefore will be, on this view, a vortex ring, the core of the vortex being made up of revolving doublets of electrons, while these may be regarded, as Larmor has suggested, as electric centres or nuclei of radial rotational strain.

252. A conception of electrons is thus seen to be capable of helping us at least to make a mental picture of the processes connected with optical and electric phenomena. We now proceed to show that it enables us to explain all those phenomena which the electro-magnetic theory of Maxwell and Hertz fails to explain.

Section II. Theory of Electric Discharge

253. Admitting that a moving electric charge will produce a magnetic field, if we further suppose that an electric current is in all cases a procession of electrons and positive electricity, we are able to construct a theory of electric discharge and to explain in a fairly simple manner the electro-dynamic action between two currents.

254. Taking one special case in which an electric discharge is passed through a De la Rive's tube at different pressures, it is found *:—

That there are several stages of the discharge: at a high pressure, the discharge is in the form of a shower or spray, consisting of an infinite number of rays. These, gradually,—as

* *Phil. Mag.* Oct. 1908.

the pressure is diminished—form into a single band or stream; as the pressure is further reduced, the band broadens and ultimately fills the whole tube as a glow-discharge.

255. Now, an electric discharge is a procession of corpuscles, shot off from the negative electrode under the influence of the electric field and ions, positive and negative, produced by collision of these corpuscles with the molecules of the enclosed gas.

The ions and the corpuscles exert electric force on one another, and as they are in motion, they exert magnetic force as well. Moreover, these masses moving through the fluid medium must exert an additional apparent force on one another, besides experiencing a viscous retardation in the direction of motion.

The resolved parts of these forces along a line of discharge affect motion and collision along this line, and therefore need not be considered further.

256. Consider now any two (parallel) streams. The resultant action due to electric and magnetic forces between two charges e, e', moving with equal velocities q, was calculated by J. J. Thomson [*Phil. Mag.* April 1881], and shown to be a repulsion of magnitude $\frac{ee'}{Kr^2}\left(1-\frac{q^2}{3V^2}\right)$, where K is the S.I.C. of the medium, r the distance between the charges, and V the velocity of light (the moving charges being assumed to be spherical).

Suppose, now, there are n positive and N negative charges in the first stream and n', N' in the second (per unit length of each stream), and let q be the velocity of a positive ion, and $-q'$, that of a corpuscle.

Then, the mutual repulsion between the elements ds, ds' will be (if e, e' be the charges of the particles in the two streams)

$$\frac{ee'}{Kr^2}\,ds\,ds'\left[nn'\left(1-\frac{q^2}{3V^2}\right)+NN'\left(1-\frac{q'^2}{3V^2}\right)\right.$$
$$\left.-nN'\left(1+\frac{qq'}{3V^2}\right)-n'N\left(1+\frac{qq'}{3V^2}\right)\right]$$
$$=\frac{ee'}{Kr^2}\,ds\,ds'\left[(n'-N')(n-N)-\frac{(nq+Nq')(n'q+N'q')}{3V^2}\right].$$

In the case of a discharge, since all the streams are similar, $n=n'$, $N=N'$, and $e=e'$.

Therefore the repulsion

$$=\frac{e^2 ds\,ds'}{Kr^2}\left[(n-N)^2-\frac{(nq+Nq')^2}{3V^2}\right] \quad \text{.........(1)},$$

the ions and corpuscles in any one stream being a self-equilibrating system.

There will certainly be repulsion, therefore, when n is large or small compared with N, for in this case the expression within square brackets becomes

$$n^2\left(1-\frac{q^2}{3V^2}\right), (N=0) \quad \text{..................(2)},$$

or

$$N^2\left(1-\frac{q'^2}{3V^2}\right), (n=0) \quad \text{..................(3)},$$

while the resultant action will certainly be an attraction of $n=N$. On this view, we may evidently argue that *in the case of a wire carrying current, $n=N$.*

257. We have already found the resultant action due to electric and magnetic forces.

We have next to consider the effect of masses moving through the fluid, supposed perfect.

If ρ = density of the fluid, a the radius of a positive ion supposed spherical, and r the distance between them; then it can be shown* that the repulsion between them, when moving with velocities q, is

$$\pi\rho\frac{a^6}{r^3}q^2.$$

If, then, there are n positive and N negative charges in either stream per unit length, and if b is the radius of a negative sphere and $-q'$, its velocity, the repulsion per unit length of either stream will be

$$\frac{\pi\rho}{r^3}\left[n^2a^6q^2+N^2b^6q'^2-2nNa^3b^3qq'\right]$$

$$=\frac{\pi\rho}{r^3}(na^3q-Nb^3q')^2 \text{......(4)}.$$

That is, the effect is always a repulsion, which decreases as the pressure of the gaseous medium decreases. The resultant

* Lamb's *Hydrodynamics*.

repulsion per unit length of either stream will, therefore, be equal to the sum of (1) and (4).

On taking account of these two sets of forces, it is easy to see that the observed phenomena must be due to repulsion and attraction between various streams of ions issuing from the various points of the cathode. While, of course, the resultant effect must depend on the relative values of N, n, q, q', the simplest supposition that we can make is that whenever these streams carry mostly one kind of ions, they will spread all round the tube, but when the streams contain an equal number of both kinds of electrified particles, and the pressure is not too high, the various streams will be attracted together and form a single stream.

Although we cannot, in the present state of our knowledge, trace in detail the change in the value of the above expression, it seems to be possible to follow the course of events in a general manner. Stated broadly, there will be repulsion between consecutive streams, both when the inside pressure is high, and also when it is low. We have thus three main stages of the discharge*.

258. It is only during the second stage that there is a steady rotation on the application of the magnetic field [in a De la Rive tube], while during the third stage the effect is that of a twist [Appendix V].

We have already seen that the variation in the number of corpuscles in the discharge at any moment must depend on the number of collisions and recombinations per second. Thus†, let

$$\lambda = \text{mean free path},$$
$$X = \text{electric field},$$
$$e = \text{charge on an ion}.$$

Then, the mean kinetic energy of an ion $= X\lambda e$.

Now, if $X\lambda e >$ a certain limiting value P, ionisation takes place.

Let $f(X\lambda e) =$ fraction of collisions that result in ionisation, i.e. $f(x) = 0$, when $x < P$.

* Recent experiments have shown that all vacuum tubes exhibit these phenomena.

† Thomson's *Conduction of Electricity through Gases* (2nd edition), Art. 229.

Also, let q = average velocity of translation of corpuscles,

N = no. of negative ions per c.c.,

then $\frac{Nq}{\lambda}$ = no. of collisions per second,

and no. of ions produced $= \frac{Nq}{\lambda} f(Xe\lambda)$.

Let $\gamma . \frac{Nq}{l}$ = no. of ions that disappear through recombination.

Then $\frac{Nq}{l}[f(Xe\lambda) - \gamma]$ = the resultant no. of ions produced, neglecting those due to positive ions.

Therefore, the equation of continuity is

$$\frac{\partial N}{\partial t} + \frac{\partial Nu}{\partial x} = \frac{Nu}{\lambda}[f(Xe\lambda) - \gamma] \quad \text{...........(5)},$$

where u = average velocity of translation due to electric field $= q$ nearly.

Now in order that there should be *steady* rotation of the discharge, *as a whole*, we must have

$$\frac{\partial N}{\partial t} = 0, \quad \frac{\partial Nu}{\partial x} = 0 \quad \text{and} \quad \therefore \quad f(Xe\lambda) - \gamma = 0.$$

But this is also the condition that the result of collision should be annulled by recombination. We conclude, therefore, that throughout the stage that this condition holds, n and N will continue to remain equal, if they are so, as we have argued [Art. 257] that they should be, at the beginning.

The quantity γ would obviously depend on the nature of the gas. The same is true of the function f. γ may also depend on the conditions of the experiment—for example, whether the discharge is intermittent or continuous. It may also depend on the pressure.

We shall presently see what experimental evidence we have on these points.

259. When the tube contains air, the curve connecting pressure and potential difference between the electrodes consists of four portions. The first portion, which is very nearly straight, corresponds to the 'spray' discharge. As the 'spray' discharge changes into the 'band' or single stream, there is a bend in the curve towards the line of pressure. The next portion is also a straight line. Gradually, the curve bends away from the line of pressure, meeting asymptotically the line of potential difference [Appendix III].

260. We have just seen that $f(Xe\lambda)-\gamma=0$ during the band or rotatory stage, that is $f\left(\frac{Xe}{p}\right)-\gamma=0$, since $\lambda \propto \frac{1}{p}$, p being the pressure.

If now γ is constant, we have $f\left(\frac{Xe}{p}\right)=\text{constant}$, so that the curve connecting p and potential difference will, in this case, be a straight line, if the spark-length is constant (as is known to be the case, both on Paschen's law and experimental grounds).

If γ is not constant, the curve will not be a straight line. This may explain why, in the case of complex gases and vapours, the curve is not a straight line.

261. If $f(Xe\lambda)-\gamma$ is negative, so that $f\left(\frac{Xe}{p}\right)-\gamma=-\alpha$, say, this will correspond to the (first) stage, during which corpuscles are decreasing as the result of collision and recombination and tending to equality with positive ions [Art. 257].

Assuming α to be constant and less than γ, we deduce that the curve connecting p and potential difference corresponding to this stage will also be a straight line, inclined at a greater angle to the line of potential difference, than the line corresponding to the rotatory stage [Art. 260].

This would explain the first bend in the curve [Art. 259].

In the third stage of the discharge, the form of the curve is consistent with the supposition that collisions are then few, and $\gamma-\alpha$ rapidly decreases with pressure [Appendix IV].

262. It has been found from experiment that the rotatory stage is entirely absent under certain conditions depending on the tube and coil. This is obviously due to the fact that for such a discharge $f(Xe\lambda)-\gamma$ is never zero, or what is the same thing, a steady state of ionisation throughout the length of the discharge tube cannot be established under the conditions of the experiment.

This is especially the case with hydrogen, and the peculiar behaviour of hydrogen in a vacuum tube, generally, would seem to be connected with this property.

That the intermittent nature of the discharge from an induction coil materially determines the value of $f(Xe\lambda) - \gamma$ is proved by the fact that when the discharge is passed from storage cells into a tube (of small spark length), the first stage—that of spray discharge—is absent.

263. It follows from experiments that during the rotatory discharge, the product of pressure and angular velocity is a constant, and proportional to the density of the gas operated on.

In order to explain this peculiarity, we observe that in the case of steady rotation, the moment of the electro-magnetic couple on the current due to the magnet is equal to the moment of the retarding forces.

The first is proportional to the current $i = \mu i$ (say), the second $= \int (A_1 + A_2)\, nur\, ds$, where

$n =$ no. of $+, -$ ions per unit length,

$r =$ distance of an ion from the axis and u the velocity,

$ds =$ an element of length of the discharge,

$A_1, A_2 =$ the retardations of $+, -$ ions per unit velocity.

Also, if q, q' be the velocities per unit electric intensity of the ions along the line of discharge, and X the electric intensity, then for the steady state

$$Xe = qXA_1, \text{ i.e. } A_1 = \frac{e}{q}; \text{ similarly, } A_2 = \frac{e}{q'}.$$

Therefore $$\mu i = \omega ne \left(\frac{1}{q} + \frac{1}{q'}\right) \int r^2 ds \text{ [Appendix V]}$$

But $$Xe(q + q')n = i;$$

therefore $$\frac{\omega}{\mu} = \frac{qq'X}{\int r^2 ds} \quad \text{.............................(6).}$$

Now, since (p being the pressure of the gas in the discharge tube)

$$p \propto \frac{1}{\lambda}; \quad \therefore\ p \propto \frac{1}{q}, \text{ and also } \propto \frac{1}{q'};$$

$$\therefore\ p\omega \propto \frac{X}{p \int r^2 ds} \text{ from (6) } \quad \text{...............(7).}$$

But $\int r^2 ds$ is found to be practically constant when the distance between the electrodes remains unchanged.

Therefore $$\frac{X}{p} \propto p\omega.$$

But $$f\left(\frac{Xe}{p}\right) = \gamma,$$

so long as there is rotation. Therefore

$$f(ep\omega) = \gamma, \text{ i.e., } p\omega = \delta, \text{ say.}$$

If γ is constant, as is practically the case for air, N_2O, etc.,

$$p\omega = \text{constant}.$$

If γ is not constant but increases with pressure, $p\omega$ will increase with pressure also.

Comparing this with experimental results, we observe that δ is proportional to molecular weight in either case.

Again, since $$V' = \int X ds,$$

where V' is the potential difference between the electrodes, if we write

$$p = aV' + b$$

(for a simple gas during the rotatory stage), a and b will be constants depending on the induction coil and the nature of the gas.

Substituting in (7), we have (assuming X to be constant along the discharge)

$$p\omega \propto \frac{V'}{p\int r^2 ds \,.\, \int ds} \propto \frac{1}{\left(a + \frac{b}{V'}\right)\int r^2 ds \,.\, \int ds}.$$

Therefore, for any given discharge tube, b is obviously small, since $p\omega =$ constant nearly, as we have just seen, $\int r^2 ds$ and $\int ds$ being taken to be constant (as we are justified in doing).

We notice, moreover, that since experiment yields the result that $p\omega$ is proportional to the E.M.F. of the induction coil, a must be proportional to this quantity also.

Finally, if the distance between the electrodes is changed, we have, taking $\int r^2 ds \propto \int ds \, (= l)$, very nearly, where l is the distance between the electrodes, $p\omega l^2 =$ constant, very nearly. This is also verified by experiment.

SECTION III. THEORIES OF DISPERSION AND ABERRATION

264. Returning to the result of Art. 239, we observe that

$-\dfrac{\partial V}{\partial x}$ = the force due to the system of electrons in volume $\int d\tau$,

= the force on the electrons other than that arising from the action of the medium + the force brought into play on account of the displacement of electrons.

If we take a small sphere of volume $\int d\tau$, the system of electrons defined by $\dfrac{\partial A}{\partial x}+\dfrac{\partial B}{\partial y}+\dfrac{\partial C}{\partial z}=-\rho$ may be replaced by a surface distribution $I\cos\theta$ per unit area of the spherical surface. And thus, as in the corresponding magnetic theory, the first part of the force in the direction of x will be $\dfrac{4}{3}\pi\dfrac{A}{k_0}$ per unit charge. For the second we may, obviously, assume an elastic force in the direction of x due to displacement of the electron, and this will be

$$-\mu(x-x_0)\,(\text{say})=-\mu\frac{A}{\rho}\,[\text{Art. }235]\ldots\ldots\ldots(1),$$

while the frictional force will be of the form $b\dot{A}$. Hence, for equilibrium,

$$\frac{\mu\int A\,d\tau}{\int\rho\,d\tau}=\frac{4\pi}{k_0}\int\rho\left(f+\frac{A}{3}\right)d\tau+\int\rho\,(\dot{y}\gamma-\dot{z}\beta)\ldots\ldots(2).$$

Putting

$$\mu=\frac{4\pi}{k_0}\mu_0,\quad \int A\,d\tau=\bar{A},\quad \int\rho f d\tau=e\bar{f},\ \text{etc.},$$

we have
$$\mu_0\bar{A}=e^2(\bar{f}+\tfrac{1}{3}\bar{A})+\frac{k_0}{4\pi}e^2(\bar{\dot{y}}\gamma-\bar{\dot{z}}\beta)\ \ldots\ldots\ldots(3),$$

or for simplicity

$$\mu_0 A=e^2(f+\tfrac{1}{3}A)+\frac{k_0}{4\pi}e^2(\dot{y}\gamma-\dot{z}\beta)\ \ldots\ldots\ldots(4),$$

all the quantities having their mean values taken over a small sphere enclosing a charge.

For motion, we have $\rho\ddot{x}=\ddot{A}$,

or
$$\int\ddot{x}\rho\,d\tau=\ddot{A}\ \ldots\ldots\ldots\ldots\ldots\ldots\ldots\ldots(5),$$

where for $\ddot{A}$, we take its mean value, as before.

Accordingly, since for a single electron of mass m, charge e and self-inductance L,

$$T = \tfrac{1}{2}(m + Le^2)(\dot{x}^2 + \dot{y}^2 + \dot{z}^2) \text{ ....(6)},$$

we have, neglecting the viscous term,

$$\frac{4\pi}{k_0}\mu_0 A + (Le^2 + m)\ddot{A} = \frac{4\pi}{k_0}e^2\left(f + \frac{A}{3}\right) + e^2(\dot{y}\gamma - \dot{z}\beta)$$
$$= \frac{4\pi}{k_0}e^2\left(f + \frac{A}{3}\right) + e(\gamma\dot{B} - \beta\dot{C}) \text{ ...(7)}.$$

Writing

$$\left.\begin{aligned} Le^2 + m &= \lambda \\ \frac{4\pi}{k_0}\mu_0 - \frac{4\pi}{k_0}\frac{e^2}{3} &= p_0^2\lambda \\ \frac{4\pi}{k_0}e^2 &= a_0\lambda \end{aligned}\right\} \text{ ....(8)},$$

we get, if the magnetic field is weak,

$$\ddot{A} + p_0^2 A = a_0 f, \text{ etc. ....(9)}$$

and $$\ddot{f} + \ddot{A} - V^2\nabla^2 f = -V^2\frac{\partial\rho}{\partial x} \text{ from (18) [Art. 238]...(10)}.$$

From equations of the type (9) we deduce, since

$$\frac{\partial A}{\partial x} + \frac{\partial B}{\partial y} + \frac{\partial C}{\partial z} = -\rho$$

and also $$\frac{\partial f}{\partial x} + \frac{\partial g}{\partial y} + \frac{\partial h}{\partial z} = \rho,$$

$$\left.\begin{aligned} \ddot{\rho} + (p_0^2 + a_0)\rho &= 0 \\ \therefore\quad \rho &= \rho_0 a_0 \cos(mt + \epsilon) \end{aligned}\right\} \text{ ....(11)}.$$

265. If $\frac{\partial\rho_0}{\partial x} = 0$, or ρ_0 is independent of coordinates, the equations will be

$$\left.\begin{aligned} \ddot{A} + p_0^2 A &= a_0 f \\ \ddot{f} + \ddot{A} - V^2\nabla^2 f &= 0 \end{aligned}\right\} \text{ ....(12)}.$$

The ordinary dispersion formulae can now be derived in the usual way.

For a plane wave (z = const.),

$$\nabla^2 f = \frac{\partial^2 f}{\partial z^2}, \text{ etc.},$$

and the solutions are

$$A = A_0 e^{ip\left(\frac{nz}{V} - t\right)}, \quad f = f_1 e^{ip\left(\frac{nz}{V} - t\right)},$$

which yield

$$n^2 = 1 + \Sigma \frac{a_k}{p_k^2 - p^2} \quad \text{.................(13)},$$

where n is the index of refraction and $\frac{2\pi}{p}$ = the periodic time of vibrations of electrons unaffected by the field, while Σ refers to the several groups of electrons that are set vibrating on account of the impressed disturbance, it being understood that

$$L_k e_k^2 + m_k = \lambda_k, \quad \frac{4\pi}{k_0}(\mu_0)_k - \frac{4\pi}{k_0}\frac{e_k^2}{3} = p_k^2 \lambda_k$$

and

$$\frac{4\pi}{k_0} e_k^2 = a_k \lambda_k,$$

corresponding to kth group of electrons.

266. Returning to equation (9), and introducing a viscous term $b_0 \dot{A}$, we get

$$\ddot{A} + b_0 \dot{A} + p_0^2 A = a_0 f, \text{ etc. ....(14)}.$$

Now, assuming the solutions

$$A = A_0 e^{ip\left(\frac{nz}{V} - t\right)}, \quad f = f_1 e^{ip\left(\frac{nz}{V} - t\right)},$$

we get

$$n^2 = 1 + \Sigma \frac{a_k}{p_k^2 - ipb_0 - p^2},$$

showing that there is absorption in this case.

If p_0 is very nearly equal to p (confining ourselves to a single electron), b_0 cannot be neglected; this further indicates that there is always absorption under these circumstances.

If p is very small, then

$$n^2 = n_0^2 + \frac{a_0}{p_0^2 - ipb_0},$$

and the real part of n is

$$n_0\left(1 + \frac{cp_0^2}{p_0^4 + p^2 b_0^2}\right),$$

where $a_0 = 2n_0^2 c$. This shows that as p increases, n diminishes. This explains anomalous dispersion.

267. It is not without interest to compare the above with the various elastic solid theories that have been proposed for the explanation of dispersion.

268. For this, let us recall the fact that in an elastic medium there is, associated with an elastic displacement, molecular rotation, and if the properties of the medium are to be capable of being expressed in terms of quantities that enter into the statement of either theory, electric displacement and magnetic force must correspond in some way with the velocity of vibration and molecular rotation. Now in the electrical theory we have two quantities, μ and K, defining the property of the medium as well as the quantities f, g, h (polarisation) and H (magnetic force), while in the theory of elasticity we have the constants σ (density), n (rigidity), and the quantities ω (molecular rotation) and ξ, η, ζ (displacement), and it will be necessary to decide upon a mode of identifying these severally.

269. Moreover, on examining the expression for energy (kinetic and potential) in terms of these two sets of quantities, it is easy to see that one mode of establishing a concordance between the two sets of phenomena is to identify (as Larmor has done) electric displacement with molecular rotation and magnetic force with ethereal velocity (in vibratory motion).

On this scheme, the electrostatic energy

$$\frac{2\pi}{K}\int (f_0^2 + g_0^2 + h_0^2)\, d\tau$$

will correspond to the energy of strain of an elastic medium. Now, since this is

$$= 2n \int (\omega_x^2 + \omega_y^2 + \omega_z^2)\, d\tau + \text{surface integrals},$$

if

$$\left(\kappa + \frac{4}{3} n\right) \Delta = 0 \text{ [Art. 191]},$$

we have (on this supposition),

$$\frac{2\pi f_0^2}{K} = 2n\omega_x^2, \text{ etc., and } \frac{1}{K\mu} = \frac{n}{\sigma} \quad \text{..........(15)}.$$

provided we further postulate the identity of the kinetic energy of a strained elastic medium, viz.

$$\tfrac{1}{2}\int \sigma\,(\dot{x}^2+\dot{y}^2+\dot{z}^2)\,d\tau,$$

and the electro-magnetic energy

$$\frac{1}{8\pi}\int \mu\,(\alpha^2+\beta^2+\gamma^2)\,d\tau,$$

where

$$\sigma = \text{density of the medium},$$

$$\xi = x - x_0,\ \text{etc.} = \text{elastic displacements of the medium},$$

$$\alpha, \beta, \gamma = \text{magnetic force},$$

$$\omega_x, \omega_y, \omega_z = \text{molecular rotation}.$$

270. From (15) we derive, provided $\frac{\mu}{\sigma} = \text{constant}$,

$$2\pi f_0^2 = \omega_x^2 \ \text{[cf. Art. 191]}.$$

And this leads to the conclusion that the resultant twist is made up of an ethereal motion in addition to an electronic displacement, neither of which is however of the nature of a pure rotation by itself.

We are now in a position to consider the equations of motion that have been proposed* to explain dispersion on a modified elastic solid theory. Boussinesq's formula is

$$m\dot{u} + M\dot{U} = \left(\kappa + \frac{4}{3}n\right)\frac{\partial\Delta}{\partial x} + n\Delta^2\xi,$$

where m is the mass, u the velocity of the ether, M, U those of 'matter' and κ = volume-elasticity, and $u = \dot{\xi}$.

271. From these we may easily derive the corresponding rotational equations, viz.

$$m\ddot{\omega}_x + M\ddot{\Omega}_x = n\nabla^2\omega_x,$$

where ω_x, etc. are the curls of ξ, η, ζ and Ω_x corresponds to ϖ_x. Boussinesq's theory is thus seen to be capable of being interpreted as based on the postulate of twists, defining the disturbed state of the medium.

* Glazebrook, *B. A. Report*, 1885.

For, if we write

$$\omega_x + \frac{M}{m}\Omega_x = f + A = f_0,$$

we get, putting $\frac{n}{m} = V^2$,

$$\ddot{f} + \ddot{A} = V^2\nabla^2 f \quad \text{.....................(16)},$$

provided

$$\nabla^2(\Omega_x - A) = 0 \quad \text{.....................(17)}.$$

Now

$$\nabla^2 A = \frac{k_0}{4\pi}\frac{\partial}{\partial x}\nabla^2\phi = \frac{k_0}{4\pi}\frac{\partial\rho}{\partial x} = 0,$$

from (11) and Art. 241.

Further, the equation of equilibrium of a material medium regarded as an elastic body would be $\nabla^2\Omega_x = 0$, etc., so that (17) amounts to the statement that in forming the equation of motion, we must regard the material medium to be at rest.

Again the material displacement is assumed by Boussinesq to be a function (F) of the ethereal displacement (ξ, η, ζ), and in particular, for dispersion, $F(\xi)$ is taken by him

$$= \lambda\xi + C\frac{\partial\Delta}{\partial x} + D\nabla^2\xi \quad \text{..................(18)},$$

where λ, C, D are constants. Hence, in our notation,

$$\Omega_x = \lambda\omega_x + D\nabla^2\omega_x, \text{ etc.} \quad \text{..............(19)}$$

or

$$A = f + D\frac{\ddot{f} + \ddot{A}}{V^2} \quad \text{.....................(20)},$$

where

$$2f = \omega_x(1 + \lambda) + \Omega_x\left(\frac{M}{m} - 1\right).$$

If we admit that f is a harmonic function, the equation (20) can obviously be written in the form

$$\ddot{A} + p_0 A = a_0 f \quad \text{.....................(21)},$$

the constants being suitably adjusted.

272. The equations of Helmholtz with the same notation as in Art. 270 are

$$\left.\begin{aligned} m\ddot{u} &= \alpha^2\nabla^2 u + \beta^2(U - u) \\ M\ddot{U} &= -\beta^2(U - u) - \alpha^2 U - \gamma^2\dot{U}\end{aligned}\right\} \quad \text{.........(22)};$$

whence

$$\left.\begin{aligned} m\ddot{\omega}_x &= \alpha^2\nabla^2\omega_x - \beta^2(\omega_x - \Omega_x) \\ M\ddot{\Omega}_x &= \beta^2(\Omega_x - \omega_x) - \alpha^2\Omega_x - \gamma^2\dot{\Omega}_x\end{aligned}\right\} \quad \text{.........(23)}.$$

On transformation, for purposes of comparison with the electron theory, i.e., putting

$$\omega_x = \lambda f + \nu A, \quad \Omega_x = \lambda' f + \nu' A \;\ldots\ldots\ldots\ldots(24),$$

we get

$$m(\lambda \ddot{f} + \nu \ddot{A}) = \alpha^2 \nabla^2 (\lambda f + \nu A) + \beta^2 \{(\lambda - \lambda') f + (\nu - \nu') A\} \;\ldots(25)$$

and $$M(\lambda' \ddot{f} + \nu' \ddot{A}) = \beta^2 [(\lambda' - \lambda) f + (\nu' - \nu) A] - \alpha^2 (\lambda' f + \nu' A) - \gamma^2 (\lambda' \dot{f} + \nu' \dot{A}) \ldots(26).$$

The equation (26) is the same as the equation (21) provided $\lambda' = 0$, while from (25) we get (if $\lambda' = 0$) an equation of the form

$$k\ddot{f} + \ddot{A} = \alpha'^2 \nabla^2 f + \beta'^2 f + \gamma'^2 A.$$

Remembering that f and A must vary as $\cos pt$ (say), we can obviously adjust the constants and variables so as to put the above equation in the form

$$\ddot{f} + \ddot{A} = \alpha''^2 \nabla^2 f,$$

which is the equation (16). [A and f differing in value from the same quantities occurring in (25) and (26), each, by a constant factor.]

Ketteler's equations are of the form

$$m\ddot{u} + MC'\ddot{U} = \alpha^2 \nabla^2 u,$$

$$MC\ddot{u} + M\ddot{U} = -\alpha^2 U - \beta^2 \dot{U}, \text{ etc.},$$

which yield as before

$$m\ddot{\omega}_x + MC'\ddot{\Omega}_x = \alpha^2 \nabla^2 \omega_x,$$

$$MC\ddot{\omega}_x + M\ddot{\Omega}_x = -\alpha^2 \Omega_x - \beta^2 \dot{\Omega}_x, \text{ etc.},$$

which are the same equations as (25) and (26), if we put

$$m\omega_x + MC'\Omega_x = f + A \quad \text{and} \quad MC\omega_x + M\Omega_x = \lambda f + \nu A.$$

On eliminating f and putting $\lambda m - MC = 0$, these yield

$$\ddot{f} + \ddot{A} = \alpha'^2 \nabla^2 f \quad \text{and} \quad K\ddot{A} + \ddot{f} = \alpha_1^2 A + \alpha_2^2 \dot{A}.$$

If f varies as $\cos pt$, this equation is of the same form as (22).

273. Inasmuch, however, as

$$\frac{\partial A}{\partial x}+\frac{\partial B}{\partial y}+\frac{\partial C}{\partial z}=-\rho$$

and

$$\frac{\partial f}{\partial x}+\frac{\partial g}{\partial y}+\frac{\partial h}{\partial z}=\rho,$$

(A, B, C) and (f, g, h) cannot be interpreted as rotations, each by itself, and to that extent these theories are less general than the electron theory.

The main difference however consists only in the fact that the elastic solid theories deal with ether and matter, while the electron theory replaces matter by electrons.

274. In attempting a comparison between the elastic solid theory and the electron theory, we have identified the electrostatic energy with the energy of (rotational) strain of the elastic medium, and the electro-magnetic energy with its kinetic energy.

But the identification is not unique, as we do not know which of the two expressions (in either system) is kinetic or which potential, or in fact whether both the energies are not (as they most likely are) kinetic. We may, therefore, if we like*, regard electric force as identical with the rate of elastic displacement and magnetic induction with molecular rotation.

On this scheme, the electro-magnetic energy

$$\frac{1}{8\pi}\int \mu\,(\alpha^2+\beta^2+\gamma^2)\,d\tau$$

is to be identified with the energy of strain

$$\int 2n\,(\omega_x{}^2+\omega_y{}^2+\omega_z{}^2)\,d\tau,$$

provided we neglect surface integrals and take

$$\left(\kappa+\frac{4}{3}n\right)\Delta=0,$$

which yields the solenoidal condition for magnetic induction, viz.,

$$\frac{\partial a}{\partial x}+\frac{\partial b}{\partial y}+\frac{\partial c}{\partial z}=0,$$

* Glazebrook's Address as President of the Physical Section, B.A., 1893.

since
$$\frac{\partial \omega_x}{\partial x} + \frac{\partial \omega_y}{\partial y} + \frac{\partial \omega_z}{\partial z} = 0$$

$\left(\text{provided } n \text{ varies as } \frac{1}{\mu}\right)$.

When we proceed to identify the kinetic energy

$$\tfrac{1}{2}\sigma \int (u^2 + v^2 + w^2)\, d\tau$$

of the elastic medium with the electrostatic energy, we observe that if the elastic medium is ethereal, the equation of condition should be

$$\tfrac{1}{2}\sigma_0 \int (u_0^2 + v_0^2 + w_0^2)\, d\tau = \frac{2\pi}{k_0} \int (f^2 + g^2 + h^2)\, d\tau;$$

while if the medium is a material medium, we must take

$$\tfrac{1}{2}\sigma \int (u^2 + v^2 + w^2)\, d\tau = \frac{2\pi}{K} \int (f_0^2 + g_0^2 + h_0^2)\, d\tau.$$

These yield the following results:

$k_0\mu = \frac{\sigma_0}{n} = \frac{1}{V_0^2}$, where V_0 is the velocity of light in the ether,

$K\mu = \frac{\sigma}{n} = \frac{1}{V^2}$ and V the velocity of light in a medium (σ, n or K, μ).

Also $\dot{\Delta}_0 \left(\equiv \frac{d\Delta_0}{dt}\right) = \frac{\partial u_0}{\partial x} + \frac{\partial v_0}{\partial y} + \frac{\partial w_0}{\partial z} = \frac{1}{k_0}\left(\frac{\partial f}{\partial x} + \frac{\partial g}{\partial y} + \frac{\partial h}{\partial z}\right) = \frac{\rho}{k_0}$,

where Δ_0 is the dilatation of the ethereal medium, while

$$\frac{\partial u}{\partial x} + \ldots + \ldots = \frac{\partial f_0}{\partial x} + \ldots + \ldots = 0,$$

i.e., the total displacement (ethereal and electronic) is solenoidal, while the volume density of electricity is proportional to ethereal expansion.

275. The various equations appropriate to an elastic medium are then found to have their exact analogues in the electron theory.

Thus the equation

$$\left(\kappa + \frac{4}{3}n\right)\frac{\partial \Delta}{\partial x} - 2n\left(\frac{\partial w_z}{\partial y} - \frac{\partial w_y}{\partial z}\right) = \sigma\ddot{\xi}$$

yields the electrostatic equation for a material medium, viz.,

$$\dot{f}_0 = V^2\left(\frac{\partial\gamma}{\partial y} - \frac{\partial\beta}{\partial z}\right), \quad \left(\text{since } \frac{\partial\Delta}{\partial x} = 0\right),$$

on *properly choosing the signs.*

The equation of motion of such a medium, viz..

$$\ddot{\xi} = V^2\nabla^2\xi,$$

yields

$$\dddot{\xi} = V^2\nabla^2\dot{\xi}$$

or

$$\ddot{f}_0 = V^2\nabla^2 f_0,$$

i.e.,

$$\ddot{f} + \ddot{A} = V^2\nabla^2(f + A) = V^2\Delta^2 f,$$

if $\nabla^2 A = 0$; while for the *free* ethereal medium we have

$$\ddot{f} = V_0^2\nabla^2 f.$$

Again, Boussinesq's equation, viz.

$$\sigma_0\ddot{u}_0 + \rho_1\dot{U} = n\nabla^2\xi_0,$$

yields, if we put

$$u_0 = f, \quad \rho_1 U = \sigma_0 A, \quad \frac{n}{\sigma_0} = V_0^2,$$

$$\ddot{f} + \ddot{A} = V_0^2\nabla^2 f.$$

The other equation of condition derivable from Boussinesq's assumed relation, viz.,

$$U = \lambda u_0 + C\frac{\partial\dot{\Delta}_0}{\partial x} + D\nabla^2 u_0,$$

gives

$$A = \lambda' f + D'\nabla^2 f$$

$$= \lambda' f + D'\frac{\ddot{f} + \ddot{A}}{V^2},$$

and, similarly, for the equations of Helmholtz and Ketteler.

276. A third method of identification will be to take

$$n(\omega_x, \omega_y, \omega_z) = (\alpha, \beta, \gamma), \quad \sigma_0(\dot{\xi}_0, \dot{\eta}_0, \dot{\zeta}_0) = (f, g, h)$$

and

$$\sigma(\dot{\xi}, \dot{\eta}, \dot{\zeta}) = (f_0, g_0, h_0),$$

giving

$$\frac{\partial\alpha}{\partial x} + \frac{\partial\beta}{\partial y} + \frac{\partial\gamma}{\partial z} = 0,$$

as in the electron theory, $\sigma_0\dot{\Delta}_0 = \rho$ and all other equations, practically the same as before.

We have, in fact, the following tabular scheme:

I

$$\omega = (f_0, g_0, h_0)$$
$$(u, v, w) = (\dot{\xi}, \dot{\eta}, \dot{\zeta}) = (\alpha, \beta, \gamma)$$
$$\left.\begin{array}{l} \sigma = \mu \\ n = K^{-1} \end{array}\right\} \text{ or } \frac{\sigma}{n} = K\mu$$

II

$$\omega = \mu(\alpha, \beta, \gamma)$$
$$(u_0, v_0, w_0) = \left(\frac{f}{k_0}, \frac{g}{k_0}, \frac{h}{k_0}\right)$$
$$\sigma_0 = k_0$$
$$n = \mu^{-1}$$

III

$$n\omega = H$$
$$\sigma_0(u_0, v_0, w_0) = (f, g, h)$$
$$\sigma(u, v, w) = (f_0, g_0, h_0)$$
$$\sigma_0 = k_0$$
$$n = \mu^{-1}$$

277. Now [Art. 197] electrical experiments lead to the conclusion that μ is very nearly constant in a dielectric but that K varies, while optical experiments make the constancy of n and variability of σ probable, so that the second mode of representation would appear at first sight to be more in accordance with the facts. It seems, however, that except in so far as the identification of the constants is concerned, the question of interaction of matter and ether is resolvable on any of these hypotheses, for we may simply have $\frac{n}{\sigma} = \frac{1}{K\mu} = V^2$, where V is the velocity of propagation, although it may not be possible to identify separately the various constants, that enter in these investigations, according to any of the schemes tabulated above. On this understanding, it does not seem to be possible to pronounce in favour of any of these in preference to the others. In spite of the uncertainty that exists in this respect, however, these various modes of representation are useful, as we have so little knowledge of the intimate nature of electric and magnetic quantities.

278. Aberration.

Take a system of moving axes carried with the movement of the earth and let ξ, η, ζ be the velocities of translation of the origin, at any time t.

Then
$$x' = x - t\xi, \text{ etc.},$$
where x, y, z are referred to the fixed axes and x', y', z' to the moving axes. And also
$$\frac{D}{Dt} = \frac{\partial}{\partial t} + \xi\frac{\partial}{\partial x} + \eta\frac{\partial}{\partial y} + \zeta\frac{\partial}{\partial z}.$$

Now, since, as we have seen, for a plane wave ($z = \text{const.}$),
$$A = A_0 e^{ip\,(nz\sqrt{k_0} - t)}$$
when the axes are fixed, we must put, when the axes are moving,
$$A = A_0 e^{ip'\,(n'z\sqrt{k_0} - t)},$$
assuming that the effect of this motion is entirely kinematical (which, however, sufficiently takes account—up to the first order—of the fact that this motion produces a magnetic field).

We have, accordingly,
$$p\,(nz\sqrt{k_0} - t) = p'\,(n'z'\sqrt{k_0} - t)$$
$$= p'\,\{n'z\sqrt{k_0} - t\,(n'\zeta\sqrt{k_0} + 1)\};$$
$$\therefore\ pn = p'n',$$
and
$$p = p'\,(1 + n'\zeta\sqrt{k_0}),$$
or
$$p = p' + pn\zeta\sqrt{k_0},$$
$$p\,(1 - n\zeta\sqrt{k_0}) = p'.$$

Modifying the equation (13), accordingly, we get, leaving out the accent wherever it occurs,
$$n^2 - 1 = \Sigma\frac{a_k}{p_k^2 - p^2}(1 - n\zeta\sqrt{k_0})^2$$
$$= \Sigma\frac{a_k}{p_k^2 - p^2}(1 - 2n\zeta\sqrt{k_0}),$$
neglecting second order terms.

If n_0 is the value of n (index of refraction) for a body at rest, we have
$$n_0^2 - 1 = \Sigma\frac{a_k}{p_k^2 - p^2},$$
whence
$$\frac{n^2 - 1}{n_0^2 - 1} = 1 - 2\zeta n\sqrt{k_0}, \text{ practically},$$

i.e.,
$$\frac{n^2 - n_0^2}{n_0^2 - 1} = -2\zeta n\sqrt{k_0}.$$

If the coefficient of retardation is ϵ, ϵ is the ratio of the difference of the two velocities of the wave (corresponding to n and n_0) to the velocity of the electron.

Or
$$\frac{1}{n\sqrt{k_0}} = \frac{1}{n_0\sqrt{k_0}} + \zeta\epsilon,$$

i.e.,
$$1 = \frac{n}{n_0} + \zeta\epsilon n\sqrt{k_0}$$
$$= \frac{n}{n_0} - \frac{\epsilon}{2}\frac{n^2 - n_0^2}{n_0^2 - 1}, \quad \text{or} \quad \epsilon = \left(1 - \frac{1}{n_0^2}\right),$$

neglecting second order terms.

Michelson and Morley's experiments show however that the second order terms are most probably zero. This calls for a modification of the above theory.

279. Without, for the present, entering on a discussion of this point, we observe that the theory does account in a fairly satisfactory manner for all optical phenomena on the postulate of certain singularities called electrons which can be subjected to experimental investigation. The difficulty of the theory is however obvious, for it leaves the nature of this singularity and that of electric charge still unexplained.

If, however, the existence of electrons or negative charges is admitted as electrified singularities in the field, Faraday tubes must be conceived to be issuing from them. These tubes, on Sir J. J. Thomson's theory, move, as we have just seen, with the velocity of light through the medium. The mode of the propagation of light or its mechanism may, therefore, be thus stated in terms of moving Faraday tubes:

280. From a source of electro-magnetic energy issue 'Faraday tubes' which move through the medium at right angles to themselves along the line of flow of energy (with the velocity of light). If the electro-magnetic energy suffers periodic variation, the number of tubes issuing will also suffer periodic variation.

This mode of representation, as Sir J. J. Thomson has remarked, recalls the emission theory developed by Newton, while the main features of the undulatory theory are conserved.

281. Again, the motion of the 'Faraday tubes,' in a medium, which may be taken to be a fluid medium of some kind, recalls the "motion of hollow vortex rings in a fluid in which portions in rotational and irrotational motion are freely mixed together and which Lord Kelvin has called vortex sponge." Now it can be shown that the equation of propagation of a plane wave in such a medium is the same as the equation of propagation of luminous vibrations in the ether and that a spiral vortex in such a medium would behave in the same way as a tube of electric force. If this view of the electro-magnetic field is accepted, the energy of the field, intrinsic and actual, must necessarily be entirely kinetic.

282. A theory of this kind, when it is well-established, will be the first step towards an adequate conception of the mechanism of the propagation of the vibratory energy which constitutes that of light. New difficulties have, however, been recently suggested which will naturally modify speculations on these lines.

283. One of these is concerned with the new idea of energy, which suggests that just as we have to postulate, not merely a sub-multiple of the Chemical atom but also an atom of Electricity, similarly, we have to admit the existence of an atom of energy [Appendix VI]; so that an oscillation giving out radiant energy must be conceived to give out always an amount, proportional to its frequency and the medium, or the recipient of the energy, to receive, necessarily, an equal amount, the constant of proportionality being a universal constant which has been called Planck's constant. This has been called by Einstein, a *light quantum* and seems to indicate that the changes involved are due to impulses. If this is indeed the case, it looks as if not merely the X-rays but all rays are propagated by pulses. This may account for many outstanding problems such as the bright line spectra of incandescent gases; but without entering into these details, it is not difficult to see that such a theory, if it proves acceptable, will profoundly modify our entire conception of the processes, operating in the electro-magnetic field. For if we must accept the operations of finite forces in effecting changes in the field, we have to admit that we can be cognisant of time, only as durations and intervals, as small as we like but not, as a continuously flowing quantity of Newton. There is of course nothing objectionable in this purely conceptual definition of absolute Mathematical time, but it would appear, on the quantum theory, that our faculties impose a limitation on its perception.

CHAPTER V

THEORY OF RELATIVITY

284. We have seen that a theory of Electrons, as singularities in an all-pervading medium is capable of explaining in a fairly satisfactory manner nearly all known optical phenomena. And thus, while the exact nature of the singularity is still a matter of speculation, it may be held to be established that it is the energy of moving electrons that is propagated as the energy of light through the ethereal medium. But just as our old conception of energy is undergoing a profound modification, an inquiry has been set on foot as to the objectivity of the ethereal medium itself, not so much as the receptacle of this energy: but as the frame of reference to which the motion of electrons has necessarily to be referred. We propose to consider this point briefly.

285. It is well known that in dealing with Dynamics of particles and of rigid bodies, so long as we confine ourselves to motion at a given position on the earth, we may as a first approximation, regard the axes as fixed in space and the resulting equations of motion are to that extent valid. Our frame of reference is taken to be fixed, although it is only fixed relatively to a moving system and so is our mode of measuring time. This is altered, if we have to compare motions at two different places, on the surface of the earth. If, however, we regard the two sets of axes as having constant velocity, with regard to each other, the equations of motion will be the same but the times will have different values—the local time and the standard time,—related to each other, in a simple manner and depending on the position of the observer, as well as on the mode of measurement.

When finally, we proceed to celestial motion, generally, a frame of reference, independent of the earth's motion is required; and the ecliptic is taken as a first approximation and a certain initial epoch, convenient to the astronomer.

286. In the same way, while for a consideration of phenomena, confined entirely to the earth's surface, the question of a frame of reference may be waived, when the electromagnetic phenomena in interstellar space have to be dealt with—in particular, the phenomenon of aberration, this becomes of primary importance. An examination of the various theories of aberration is, therefore, necessary, in order that we should be able properly to appraise its significance.

287. On the emission or corpuscular theory of light, the direction in which a star is seen is the direction in which corpuscles, supposed to be shot off from the star appear to come to the observer, partaking of the earth's motion. Actual calculation shows that this exactly accounts for the displacements observed—in a non-dispersive medium, such as the interstellar space is, but, even within these limits, the explanation has no greater justification than the corpuscular theory itself.

288. On the Undulatory theory, the explanation naturally presents a difficulty, as the operation of the medium has necessarily to be taken into account. An explanation on the elastic solid theory is possible, as we have seen (Art. 103), provided we assume the operation of body-forces, proportional to the second differentials of displacements with regard to time. But this is a mere matter of analysis, which requires further physical interpretation. Finally, the explanation offered by Fresnel (Art. 216) gave also the correct result, but only up to the *first order*. We may recall here, that it proceeds on the hypothesis that a portion of the ether in the space occupied by the moving body is bound with the body and is carried with it with the velocity it possesses, the rest being free ether and stationary, as in free space. Accordingly, when a ponderable medium is in motion, the velocity of light in it is increased by a fraction (the so-called dragging coefficient of Fresnel) of the velocity of the medium, a conclusion which the experiment of Fizeau amply verified.

289. Fizeau's experimental device* was to divide a beam of light into two portions, which passed along two parallel tubes which contained water, at rest or flowing in two opposite directions, with the same velocity and then cause them to interfere. The shifting of interference fringes, with stationary and flowing water, enabled the dragging coefficient to be calculated, and it was found to agree with Fresnel's theoretical value.

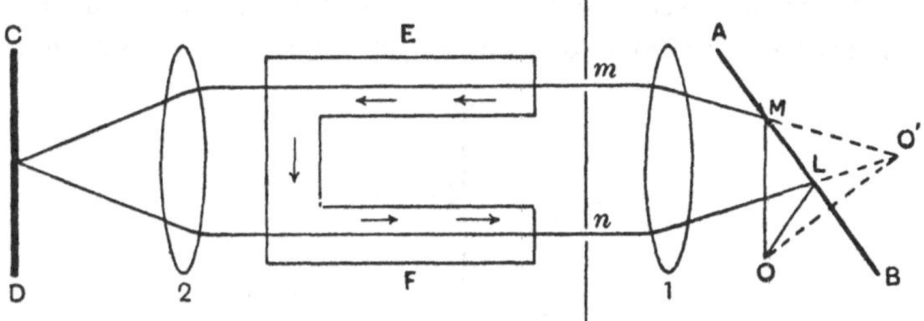

O is a source of light, *AB* a plate of glass, at which rays *OL*, *OM* are reflected. These rays after passing through the lens (1), proceed through apertures *m*, *n* in a screen, to the tubes *E* and *F* containing water and thence through the lens (2) to the mirror *CD*.

The rays, after reflection at *CD*, interchange their paths and ultimately passing through the first lens and the plane glass *AB*, produce interference fringes.

290. If, ϵ is the dragging coefficient, $\epsilon v =$ ether-drift, where v is the velocity of the moving medium. Hence we have, if $V =$ velocity of light in the quiescent medium, relatively to the earth, $V \pm \epsilon v =$ velocity of light in the moving medium, according as v is in the direction of propagation of light or opposite to it.

Accordingly, if l is the length of the total path of each beam in Fizeau's experiment, the retardation of one beam relative to the other is

$$\frac{l}{V-\epsilon v} - \frac{l}{V+\epsilon v} = \frac{2l\epsilon v}{V^2},$$

neglecting square terms.

Thus, there will be displacement of the interference bands; the displacement depending on the velocity of the medium; for we may take $V = V_0 + u$, where u is the velocity of the earth relative to the ether, resolved in the direction of propagation and $V_0 =$ velocity of light *in vacuo.*

* Fizeau, *Comptes Rendus*, vol. XXXIII, 1851. The apparatus is not figured in the paper but the description is sufficiently explicit.

291. Thus, the effect in Fizeau's experiments (depending as it did on the velocity of the medium, namely, water) gave no information as to the velocity of the earth (relative to the quiescent ethereal medium of Fresnel). Let, now, the experiment be slightly modified and instead of two parallel tubes, two tubes at right angles to each other be taken with their directions, along and perpendicular to the direction of the earth's velocity (AD, say); then by a suitable arrangement of mirrors (see fig.), the two beams may be finally made to interfere,

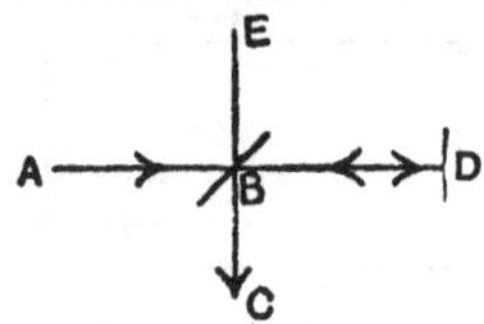

after they have traversed the tubes. If, now, the whole apparatus be rotated through 90°, the resulting displacement of the fringes, if any, will depend on the square of earth's velocity, relative to the ether.

Since in this experiment, light travels in air, $\epsilon = 0$, so that we have only to take account of the earth's velocity, u, relative to the ether.

If $$BD = l_1, \text{ and } BE = l_2,$$
time from B to D and back in the first experiment

$$= \frac{l_1}{V_0 + u} + \frac{l_1}{V_0 - u};$$

also time from B to E and back $= \dfrac{2l_2}{V}$, where V is the velocity of light *relative* to the earth in the direction BE, i.e.,

$$V^2 + u^2 = V_0^2.$$

Hence the retardation producing interference

$$= \left[\frac{l_1}{V_0 + u} + \frac{l_1}{V_0 - u} - \frac{2l_2}{\sqrt{V_0^2 - u^2}}\right] \quad \text{...........(1)}.$$

Similarly, in the second experiment, the retardation

$$= \left[\frac{2l_1}{\sqrt{V_0^2 - u^2}} - \frac{l_2}{V_0 + u} - \frac{l_2}{V_0 - u}\right] \quad \text{...........(2)}.$$

The difference between (1) and (2) producing displacement of the fringes is $\left(l_1 + l_2\right) \frac{u^2}{V_0^2}$, neglecting higher powers.

292. This is the principle of Michelson and Morley's celebrated experiment but no such effect has actually been observed, although the accuracy of the experimental arrangement was such that it could not fail to be detected, if it existed at all. This being so, the conclusion is either that the ether of space is moving with the velocity of the earth, which is evidently untenable, or that there must be some compensating cause which leads to the null-effect observed. If, for instance, as was suggested by Fitzgerald, the tubes contract when rotated in the direction of motion by a suitable amount, a contraction which will again require explanation in its turn, the null-effect is explained*; but if this is the case, it will follow that it will not be possible, by means of any such experiment to find the velocity of the earth relative to the ether, if it exists. Moreover, experiments show that no optical, electrical or mechanical effects are observable that can be ascribed to the contraction, required by Fitzgerald's hypothesis, so that there must be causes nullifying the effect of this contraction, as deep-seated as the mechanism of these phenomena themselves.

293. We have seen that neither Maxwell's simple theory nor Hertz's modification of it for a moving medium is competent to account for aberration, even up to first order effects, as observed in Fiseau's experiments; but that, with the help of the Electron theory, the Fresnel formula can be obtained on the simple postulate that the total differential with regard to time, involves the velocity of the moving medium, as in Hydrodynamics. This, however, as we have seen does not explain the null-effect of Michelson and Morley's experiments.

* For this, we must have

$$\frac{l_1 V_0}{V_0^2 - u^2} - \frac{l_2}{\sqrt{V_0^2 - u^2}} = -\frac{l_2' V_0}{V_0^2 - u^2} + \frac{l_1'}{\sqrt{V_0^2 - u^2}},$$

where l_1' and l_2' are the changed lengths of l_1 and l_2 on Fitzgerald's hypothesis. Hence, it is only necessary and sufficient, that

$$\frac{l_1}{l_1'} = \frac{l_2'}{l_2} = \left(1 - \frac{u^2}{V_0^2}\right)^{\frac{1}{2}}.$$

294. Now we observe that in deriving the dragging coefficient of Fresnel, on the basis of the Electron theory, we have varied the geometrical co-ordinates only but kept t unchanged [Art. 289]. It will be recalled, however, that the effects due to a disturbance (luminous or otherwise), at any epoch do actually depend on a preceding epoch [Art. 195]. And it is, accordingly, convenient to distinguish between local time and a standard time (as in Astronomy) in respect of the propagation of any such disturbance. It would thus appear, *a fortiori*, that if the medium is in motion, the epoch from which time is to be reckoned should require to be specified, and we should, thus, in dealing with electro-magnetic phenomena in a moving medium, have to distinguish between two kinds of time. If this is granted, then, if the motion is in the direction of Z, the two times will be connected by an equation of the form,

$$t' = t - \lambda z.$$

295. This being premised, the null-effect above referred to may be explained, if we adopt the procedure of Lorentz and Larmor. This is to transform the equations of the electro-magnetic field, in terms of new co-ordinates, referred to moving origin: the new x', y', z', t' being related to the old x, y, z, t, so that, after transformation, the form of the differential equation $\ddot{\phi} - \frac{1}{V_0^2}\nabla^2\phi = 0$ remains unchanged. This amounts to the assumption that the charge of an electron and the associated magnetic effects are invariants for all such transformations. It is then found that the Fitzgerald contraction can be explained as well as the various null-effects, already referred to.

296. Now for a moving medium, $(0, 0, \zeta)$ [Art. 278] we have the formulae of transformation,

$$x' = x, \quad y' = y, \quad z' = z - \zeta t \quad \text{..............(3).}$$

If we introduce the further equation, $t' = t - \lambda z$ it is found that on transforming the expression,

$$\ddot{\phi} - V_0^2\nabla^2\phi,$$

it cannot be made to retain its original form, by any choice of λ.

297. In order to meet this difficulty, let us modify the formula (3) (on the basis of Fitzgerald's hypothesis) by taking

$$z' = \mu(z - \zeta t) \quad \text{........................(4),}$$

and seek to find, if possible, the values of λ and μ so as to secure the formal invariance of the differential equation in ϕ. This will require that the equation connecting t and t' will have to be modified also, and as there are three conditions to be satisfied, we should take $$t' = \lambda t - \nu z.$$

Now, since
$$\frac{\partial}{\partial z} = \mu \frac{\partial}{\partial z'} - \nu \frac{\partial}{\partial t'} \quad \text{.....................(5)},$$
and
$$\frac{\partial}{\partial t} = -\mu \zeta \frac{\partial}{\partial z'} + \lambda \frac{\partial}{\partial t'} \quad \text{.....................(6)},$$
we have
$$V_0^2 \frac{\partial^2 \phi}{\partial z^2} - \ddot{\phi} \equiv V_0^2 \left[\mu \frac{\partial}{\partial z'} - \nu \frac{\partial}{\partial t'}\right]^2 \phi - \left(\lambda \frac{\partial}{\partial t'} - \mu\zeta \frac{\partial}{\partial z'}\right)^2 \phi$$
from (5) and (6).

If this is to reduce to the form
$$\left(V_0^2 \frac{\partial^2}{\partial z'^2} - \frac{\partial^2}{\partial t'^2}\right) \phi,$$
we must have
$$V_0^2 \mu\nu = \lambda\mu\zeta, \text{ i.e. } \nu = \frac{\lambda\zeta}{V_0^2} \quad \text{...............(7)},$$
$$\frac{V_0^2 \mu^2 - \mu^2 \zeta^2}{V_0^2} = \frac{V_0^2 \nu^2 - \lambda^2}{-1} = 1 \quad \text{...............(8)}.$$
$$\therefore \lambda = \mu = \frac{1}{\left(1 - \frac{\zeta^2}{V_0^2}\right)^{\frac{1}{2}}} \quad \text{...... from (7) and (8)},$$

giving the magnitude of the contraction, required on Fitzgerald's hypothesis. But this completely explains the non-existence of the second order terms in the Michelson-Morley experiment. For, since the distance between two points x_1, x_2 in stationary medium, is $x_1 - x_2$ the new distance becomes
$$x_1' - x_2' = \mu (x_1 - x_2) = \frac{x_1 - x_2}{\sqrt{1 - \frac{\zeta^2}{V_0^2}}} \quad \text{...........(9)},$$
as required on that hypothesis [Art. 291, note].

298. From the above formulae of transformation, it follows that
$$dx'^2 + dy'^2 + dz' - V_0^2 dt'^2 = dx^2 + dy^2 + dz^2 - V_0^2 dt^2 \quad \text{......(10)},$$
which means that the velocity of light is the same in both x, y, z, t, and x', y', z', t', systems.

299. The formal invariance of the differential equation in ϕ [Art. 294] which also involves the constancy of V (further emphasized by the equation (10)), have been interpreted by Einstein in an altogether different and novel manner*. In doing so, he is led to a new principle which he calls the principle of relativity.

300. We have seen that the null-effect of Michelson and Morley's experiment can only mean that the contraction of the tube used in that experiment is independent of the material used and is automatic. It necessarily follows, therefore, that it will not be possible to detect this contraction, except as a matter of indirect deduction. Now, it is well known that gravity as an universal property of matter is not detectable directly, as all bodies are equally affected, except by a special device. Here, the effect is more deep-seated still, as we fail to detect it by any device whatsoever. And this stands to reason, as it depends on the velocity of the body relative to our assumed frame of reference which, it appears, is a quantity that cannot be directly measured. We may, in fact, regard the frame of reference to move with any arbitrary velocity that we choose. This would seem at first sight to lead to the conclusion that we are precluded from discovering any physical laws, whatsoever. Since, however, such an intellectual barrenness is not in accordance with our experience, it seems to be reasonable to admit with Einstein that: "The laws of physical phenomena are the same, whether these phenomena are referred to the system of co-ordinates in any frame of reference or any other system of co-ordinates moving uniformly with respect to it, with any arbitrary velocity whatsoever." Now, as we have already seen, the ordinary equations of motion remain the same, whether the axes are fixed or are moving uniformly, so that, stated in this form, the principle appears to be by no means new. But there is a special implication, contained in the principle, viz. that time in the moving system depends on the time as well as the corresponding space co-ordinates of the first system, as in the Lorentz-Larmor theory. In seeking now the [additional] principle on which the relation

* See a highly suggestive discussion of this point by Whitehead, *Principles of Natural Knowledge*, Ch. III.

between the two times is to be determined, we have to bear in mind that if these phenomena, as being conditioned by time and space, are to be capable of scientific treatment, there must be some connecting link between any two frames of references, moving with any arbitrary velocity, with respect to each other. This is supplied by the additional postulate: namely, that the velocity of light is the same, whatever be the frame of reference (whether the source is at rest or in motion).

301. If we adopt the view that the relations between x, y, z, t and x', y', z', t' are linear* and of the form, say,

$$x' = \kappa x, \quad y' = \kappa y, \quad z' = \lambda(z - \zeta t)$$

and $$t' = lx + my + nz + pt \qquad\qquad (11),$$

κ, λ and p being (as yet) undetermined multipliers, it can be shown that if the velocity of light is to be the same in the two systems, i.e. if

$$\frac{x^2 + y^2 + z^2}{t^2} = \frac{x'^2 + y'^2 + z'^2}{t'^2} = V_0^2 \qquad\qquad (12),$$

we must have $$l = 0, \quad m = 0, \quad n = -\frac{\lambda\zeta}{V_0^2},$$

$$\lambda = \kappa\left(1 - \frac{\zeta^2}{V_0^2}\right)^{-\frac{1}{2}} = p \qquad\qquad (13)\dagger.$$

Now, writing $\kappa = 1$, we get the formulae of the Lorentz-Larmor transformation; and the Fitzgerald contraction as well as other null-effects are then similarly explained.

302. It is interesting to note that the transformation

$$x' = x, \quad y' = y, \quad z' = \mu(z - \zeta t), \quad t' = \mu\left(t - \frac{\zeta z}{V_0^2}\right) \qquad (14)$$

leads to a simple deduction of Fresnel's dragging coefficient.

* On the principle that a plane wave should remain plane even after transformation.

† In fact, since

$$\kappa^2(x^2 + y^2) + \lambda^2(z - \zeta t)^2 - V_0^2(lx + my + nz + pt)^2 \equiv \kappa'^2(x^2 + y^2 + z^2 - V_0 t^2)$$

on comparing co-efficients, we get the above results.

For,
$$dz' = \mu(dz - \zeta dt),$$
$$dt' = \mu\left(dt - \zeta \frac{dz}{V_0^2}\right), \text{ from (14).}$$
$$\therefore \frac{dz'}{dt'} \equiv \zeta' = \frac{\dot{z} - \zeta}{1 - \frac{\zeta \dot{z}}{V_0^2}}, \text{ i.e. } \dot{z} = \frac{\zeta + \zeta'}{1 + \frac{\zeta\zeta'}{V_0^2}},$$

but ζ' = velocity of light to an observer moving with the medium, and

ζ = velocity of the medium to a second observer,

$\therefore$ $\dot{z}$ = velocity of light to the second observer

$$= (\zeta + \zeta')\left(1 - \frac{\zeta\zeta'}{V_0^2}\right)$$
$$= \zeta' + \zeta\left(1 - \frac{\zeta'^2}{V_0^2}\right), \text{ neglecting } \frac{\zeta^2\zeta'}{V_0^2},$$
$$= \zeta' + \zeta\left(1 - \frac{1}{n^2}\right)$$

if n is the index of refraction.

303. We have already adverted to the simple case of local time which depends on the standard time as well as the position of the observer. Of this, accordingly, Einstein's conclusion may be said to be a generalisation, limited by the further proviso as to the constancy of light-velocity. With regard to the latter principle, it should be borne in mind that all our observations, celestial as well as terrestrial, are ultimately based on an optical method, and all our standards depend on the velocity of light. Since, therefore, any uncertainty in this introduces complete uncertainty into the entire range of our experience, Einstein's postulate of constant light-velocity, which also enters as a constant in Lorentz and Larmor's equations, seems to be, *a priori*, justified.

304. We may regard the interesting feature of the present view in another manner*. Suppose we assume the earth's velocity relative to the ether to be as much as 161,000 miles per second in a vertical direction. Then, a rod six feet long, when horizontal, contracts to 3 feet, when placed vertically. But the

* Eddington, *Report on the Theory of Relativity.*

standard yard-measure will change in the same way, so that it will not be possible to notice the change. Nor will anything be observable, as the rod is rotated from the horizontal to the vertical position, if we admit that the image in the retina changes *pari-passu*, which is apparently the case. We may apply other tests, electrical, optical, etc. But they all fail, so that the practical result is unchanged. What is changed, however, is the point of view. We begin to regard the intimate nature of phenomena from a new aspect. We may even analyse this aspect further. When a rod is set moving uniformly, we say it contracts, but the contraction is only a way of describing the new spatial relation that comes to subsist between the rod and the observer, the contraction being only determinable with reference to the particular observer, being different for different observers. If then the spatial relations are different for different observers, we reach the remarkable conclusion that each observer carries his own (Four dimensional*) space with him†.

* Cf. Minkowski; since t enters in the differential equation for ϕ practically in the same way as x, y, z. Moreover, the idea of space really involves the idea of transference, i.e. of change of position *in time*, so that the idea of time is intimately associated with that of space. In fact, time enters so fundamentally into the measurement of space that it may be regarded as a characteristic property of space, in the abstract, in the same way as its extension proper in three directions. By an obvious extension of meaning of the word dimensions, therefore, we may regard time as a dimension of space and since all physical phenomena are conditioned by space and time, we may describe these as occurring in four dimensional space.

† Prof. Karl Pearson (*Grammar of Science*) speaks of an unfortunately common form of emotional science which revels in contrasting the infinities of space with the finite capacities of man. He argues that the space of our perception—the space in which we discriminate phenomena, is exactly commensurable with the contents of that finite capacity, which we term our perceptive faculty, so that the only infinite space we know of is a product of our own reasoning faculty. The mystery of space, according to Karl Pearson, whether it be the finite space of perception or the infinite space of conception, has indeed no existence outside each human consciousness. The theory of relativity suggests, however, that space possesses properties which are neither entirely perceptual nor entirely conceptual but which partake of the properties of both at the same time. In the same way, the old formula that the property of a body, extension for instance, is either in the body or in the consciousness of the observer is seen to be inadequate. For, after all, it appears that it depends on both, so that the distinction between the perceptual and the conceptual on the one hand and that between the subjective and the objective on the other seem to tend to be obliterated.

305. The justification of a theory is often* measured by its power to explain outstanding problems and, accordingly, an attempt has been made to apply the theory of relativity to explain the most celebrated of these, viz., the intimate nature and law of gravitation. When this is done, it appears that a profound modification of the Newtonian theory is called for. And it is, moreover, found that when this modified theory is applied to specific problems, the results are singularly satisfactory.

One of the most successful of these applications is to a well-known problem in planetary theory, namely, the known discrepancy between the observed period of rotation of the orbit of mercury (574 seconds per century) and the calculated amount, on the Newtonian theory of perturbations, due to the action of the other planets (about 532 seconds). This can be completely accounted for on the theory of relativity and, although it has been argued, on the other side, that a possible modification of the ordinary theory is competent to yield the same result, it appears that a provisional pronouncement on the point should be in favour of Einstein's theory.

But a more striking verification† of the new theory of gravitation based on the principle of relativity is now forthcoming. Einstein had predicted that rays of light would suffer deviation in a gravitational field. But if this is to be the case, rays from stars, which have to pass through regions in the neighbourhood of the sun, in reaching the observer should suffer a known deviation on account of the sun's gravitational field. In fact, rays from a star seen close to the limb of the sun should experience a total deflection of 1·74″. This prediction of theory has been amply verified from observations made at the last total eclipse of the sun‡. The theory of relativity may, therefore, claim for the present to give an insight into the nature of space and time, which is new to science.

This being so, it suggests a difficulty, which though serious does not seem to be insurmountable. For it seems to indicate that any two frames of reference are equally valid frames of

* In our ignorance of the ultimate nature of things.

† Not a confirmation, until we can establish the *uniqueness of the solution.*

‡ *A Determination of the Deflection of Light by the Sun's Gravitational Field,* by Sir F. Dyson, Prof. Eddington and C. Davidson.

reference, no matter how they are moving relatively to each other, provided the unique condition as to the constancy of light velocity is satisfied, with reference to them. If one of these frames of reference is situated in the ethereal medium, supposed to be the same as the electromagnetic field, this apparently leads to the conclusion that the ether may be supposed to be moving with any arbitrary velocity, whatsoever.

Now various lines of arguments seem already to point to the conclusion that the electromagnetic field, if identified with an ethereal medium cannot be held altogether to be either inert or immobile. There is the intrinsic [kinetic] energy of the field to be accounted for, as well as the property of the medium as a carrier of momentum of radiation. And as to the arbitrary nature of the postulated velocity, since such a medium is necessarily also a field for other phenomena, besides electromagnetic, such as gravitation, it does not appear that the theory of relativity will dispose of the physical existence of the *ethereal model,* until a better one can be found, which shall explain the intimate nature of the various concepts of modern physics, corpuscles and positive particles, electric charge and magnetic force, gross matter and gravitation, in one comprehensive scheme.

CHAPTER VI

SUMMARY. CONCLUSIONS

A Brief Outline of the Optical Theories

306. We have seen that the various attempts to understand the mechanism of the propagation of light and the nature of the luminiferous medium have been so far only partially successful. They help us, however, to make a mental picture of the intimate nature of optical phenomena and suggest the direction in which we must seek for the elucidation of the mysteries connected with them. It will, therefore, be not without interest to review the present position, free as far as possible from technicalities.

307. To the question—what is light?—the obvious answer, that will be readily acceptable to all, is that it is a form of energy. When, however, the nature of this energy is investigated, the inquiry is found to be beset with difficulties. Apart from the view that all energy is really kinetic, we may safely argue that optical energy must be associated with motion of some kind. But, is the motion translatory, rotational or vibratory or is it a suitable combination of these? Now, it is obvious that this energy must be associated, in some way, with a motion of translation, for light takes time to reach the eye from the source. But when this is postulated, the inquiry only acquires further complexity. We have the source, the eye receiving the sensation and the intervening space. In what form does the energy exist in all these and what is the nature of the medium which takes part in the transfer?

308. We may imagine highly attenuated particles or corpuscles shot off with a certain very great velocity and, proceeding through space, to penetrate the eye and thus cause the sensation of vision. No one, at the present day, will hold that there is anything absurd in this, *a priori,* in view of the remarkable properties of radium and the negative rays. But, at one time, to base any scientific theory on the postulate of imponderables appeared to physicists, *ipso facto,* to argue mediaeval mysticism. It is, rather, because such a theory does not enable us to completely explain the various phenomena connected with light, in particular, the periodic character of optical phenomena and polarisation, that we have to reject it, but even then, only partially.

309. Next, let us consider vibratory motion. Such a motion, alone, involves periodicity and we are thus driven necessarily to postulate a vibratory motion as essentially associated with the energy of light. Transfer of this energy, therefore, must take place as a transfer of vibratory motion—in free space or in ordinary matter. This necessarily leads to the inquiry as to what it is that actually vibrates, and arising out of it, we have to answer the equally pertinent question, as to how the transfer does take place—by a bodily transfer of the vibratory medium or free space, or by a point-to-point transfer.

310. But we have not yet exhausted the various known categories of motion. Admitting that optical energy is vibratory and that its transference involves a motion of translation, we may still ask the question—Is there anything in the nature of rotation going on, as a part of the intimate processes connected with optical phenomena? Now, the phenomena of polarisation show that the vibratory motion which gives rise to, or is associated with, the energy of light must be transverse to the direction of propagation—the direction of the ray. This would lead to a tentative answer to the question in the affirmative, for which stronger reasons will be presently forthcoming.

311. Going back now to the questions proposed in a previous paragraph [Art. 309], we may imagine a bodily transfer of a material something, which one finds it difficult to realize, in view specially of the enormous velocity of transfer, or on the analogy of sound, consider the propagation to be a point-to-point transfer of motion and this necessarily leads to the conclusion that the 'free' space is a plenum—full of something which we may call (provisionally) the ether and which can take up and transmit vibratory energy of extremely minute periodicity. This being granted, the fundamental phenomena of reflection, refraction, interference (including diffraction) and rectilinear propagation of light find a simple and readily intelligible kinematical explanation.

312. But if the medium is capable of taking up and transferring vibratory motion, it must possess a property akin to elasticity; for the only—at any rate the simplest—way we can conceive such a transfer of motion to take place is by the transfer of the elastic deformation of successive portions of the medium, as in the case of sound.

313. The inquiry is thus naturally concentrated on the nature of the elastic property and inertia of the medium. And it is easily seen that it must be taken to possess properties akin to that of an elastic solid; for optical vibrations, being transverse, the medium should be capable of undergoing transverse or shearing strain.

314. Now, the elastic properties of a solid body are its compressibility and rigidity. When any portion of such a body is slightly deformed and let go, two types of vibrations are set up and two types of waves are propagated in the body, one dilatational, on account of its compressibility, and the other transverse, due to its rigidity. Moreover, since the elastic deformation of a solid may be conceived to be made up of a linear displacement and a rotation of each element of volume (molecular rotation), the transverse wave must on this view be connected with the propagation of molecular rotation of the medium. If, then, the medium that takes part in the propagation of light does so on account of its elasticity, there should be two sets of waves. But only one of these, the torsional, can be identified as light waves; and in this event, its rigidity divided by its density should be equal to the square of the velocity of propagation of the transverse wave, or the velocity of light in the medium. There remains, then, the other, the dilatational wave, to account for.

315. But, before we proceed to consider this point, we are confronted with an initial difficulty, the difficulty of conceiving the interstellar space, filled with a substance, possessing the properties of rigidity and incompressibility. For this, however, one has only to note that the medium may be ordinarily absolutely devoid of rigidity and be perfectly compressible; when however a very rapid succession of vibratory motions is imposed on it, it may behave as a rigid body or appear to be incompressible, just as, although the atmospheric air is practically perfectly compressible in respect of pressure exerted in finite time, the rapid vibratory motion of a tuning fork induces in it a certain incompressibility. The elasticity of the ethereal medium is, therefore, its quasi-elasticity, quasi-incompressibility and quasi-rigidity, which does not contradict any of the canons of common sense or of common experience.

316. This difficulty being thus (provisionally) got over, we have to address ourselves to the question of the dilatational wave. For this, we may frame alternative hypotheses. We may suppose the dilatational wave either non-existent or propagated with very great (infinite) velocity. The two hypotheses* however amount practically to the same thing—if we frame the additional hypothesis that dilatational waves, if they exist, do not give rise to the impression of light. We may note, in passing, that a dilatational wave in a practically incompressible ether may explain gravitation.

317. When we proceed to consider the phenomena occurring at the surface of separation of two media, we have a considerable choice of alternatives; we may consider (1) the density of

* If the velocity of propagation is to be infinite, the ether must be held to be incompressible, while if the dilatational wave is to be non-existent, the ether must possess certain peculiar ('labile') properties. In the latter case, the medium must be assumed to be so constrained as to have no velocity at the bounding surface, while a certain relation must hold between the elastic constants of the medium.

ether the same, (2) the rigidity the same, or (3) postulate various combinations of surface conditions, according as we regard the ether as incompressible or 'labile.'

318. A further complication arises when we proceed to consider double refraction of uniaxial or biaxial crystals. It is, *a priori*, evident that the properties of the medium will be related to the axes of crystalline symmetry in these crystals. But this being admitted, a variety of hypotheses are seen to be possible:

(1) Density may depend on direction.

(2) The elastic constant may depend on direction.

(3) Both density and elastic constants may depend on direction.

(4) The ether may be incompressible or labile.

319. Connected with this inquiry is the further one as to the influence of matter on the medium:—How and how far it affects (1), (2), (3)? In this regard, we can only try to work out the conclusions, derivable from the various hypotheses, and compare these conclusions with the results of experiments. As the result of such a course of inquiry, we conclude that the hypothesis of a labile ether with isotropic elasticity but æleotropic inertia, with suitable constraints imposed on the free motion of the ether, due to the action of matter, most nearly represents the facts of experiments, but this leads to the conclusion that the transverse vibrations constituting light are perpendicular to the plane of polarisation and perpendicular to the ray, and do not lie in the wave front. The constraints to be imposed, moreover, depend on the particular structure of the medium, about which we can only speculate. It is not difficult, moreover, to see that the dependence of density on direction postulated in this form of the theory must be due to the action of matter.

320. The alternative hypothesis (MacCullagh's) of isotropic inertia and æleotropic elasticity, depending on molecular rotation only, yields also a fairly satisfactory explanation, on the understanding that vibrations are in the plane of polarisation, provided we hypothecate further that there is a certain intrinsic molecular strain, which alone can be consistent with such elasticity, in a medium satisfying the criterion of stability.

321. On the whole, therefore, the attempt to construct a model of the ether by enduing it with elastic property has only met with a limited success. Of this we are sure, however, that in the medium a transfer of energy is going on by means of a strained condition which is continuously changing.

322. Now, in an electro-magnetic field also, there is a transfer of energy continuously going on, whenever an electromotive force acts in the field, producing a strained condition throughout the field. The phenomena of induction (electrostatic and electro-magnetic), indeed, make this view of the field probable, of which the experiments of Hertz demonstrate the correctness. It is, moreover, found that the velocity of propagation of a periodic electro-magnetic disturbance is equal to the velocity of light. We are assured, therefore, that there is an intimate connection between optical energy and the energy of periodic electro-magnetic disturbance, that, in fact, the two must be identical, and the processes that are going on in the electro-magnetic field are identical with those going on in the luminiferous medium, so that the luminiferous medium is really the same as the electro-magnetic field.

323. When this is recognized, the simpler optical phenomena (reflection, refraction and double refraction) find a direct and coherent explanation on the basis of Maxwell's equations of the electro-magnetic field. In the electro-magnetic field, moreover (under periodic electromotive force), we have periodic 'electric displacement' and periodic magnetic force, perpendicular to each other and both perpendicular to the direction of propagation. Now, the previous theories left the question of the direction of vibration, in its relation to the direction of propagation, open. The electro-magnetic theory supplies the answer, for according to it, there is disturbance in both azimuths.

324. All this is satisfactory so far. But the success of the electro-magnetic theory of Maxwell is by no means complete.

It fails to explain aberration, metallic reflection and dispersion. And, even apart from this, it labours under the objection that the explanation of optical phenomena it supplies, is not only not dynamical, but is stated in terms of quantities (electricity and magnetism), of which the nature is unknown and will require further investigation.

325. This is by no means easy. As a first step towards this investigation, we may compare it with the elastic solid theory. In view of this, we note that in an elastic medium there is, associated with an elastic displacement, molecular rotation, and if the properties of the medium are to be capable of being expressed in terms of quantities that enter into the statement of either theory, electric displacement and magnetic force must correspond, in some way, with the velocity of vibration and molecular rotation. Here, again, considerable choice is possible. In the electro-magnetic theory, we have the two quantities defining the property of the medium μ (magnetic permeability) and K (specific inductive capacity), as well as the quantities, polarisation and magnetic force, while in the theory of elasticity, we have the constants of the medium, the density (σ) and rigidity (n), as well as the quantities, molecular rotation and displacement; and it is natural to inquire if these two sets of quantities can be expressed in terms of each other, at all.

326. Now, on examining the expression for energy (kinetic and potential) in terms of these two sets of quantities, we find that one way of establishing a concordance between the two classes of phenomena would be to identify electric displacement with molecular rotation and magnetic force with ethereal velocity (in vibratory motion). But the identification is not unique, as we do not know which of the two expressions (in either system) is kinetic or which potential, or in fact whether both the energies are not (as they most likely are) kinetic. We may therefore, if we like, regard electric force as identical with the rate of elastic displacement and magnetic force with molecular rotation. On the first mode of identification, it is found that the magnetic permeability must be taken to be proportional to the density of the elastic medium, and specific inductive capacity, to the inverse of rigidity, while on the second (as well as on a third mode of identification) density is to be taken to be proportional to specific inductive capacity and rigidity inversely proportional to magnetic permeability.

327. Now electrical experiments lead to the conclusion that μ is very nearly constant in a dielectric but that K varies, while optical experiments make the constancy of n and variability of σ probable, so that the second mode of representation would appear at first sight to be more in accordance with the facts. But if we regard the identification as having reference to free ether only, either mode of representation might be regarded as provisionally fitting in with facts, the question of inter-action of matter and ether still remaining open. It is further conceivable that we may have $\frac{\sigma}{n} = K\mu = V^{-2}$, where V is the velocity of propagation, although it may not be possible to identify separately the various quantities that enter in these investigations, according to any of the schemes specified above. In spite of the uncertainty that exists in this respect, this mode of representation is useful, as we have so little knowledge of the intimate nature of the electric and magnetic quantities.

328. But a further analysis is possible. Regarding the ether to be a fluid medium, we may naturally seek to identify the various known hydrodynamic quantities with quantities that enter in the discussion of electro-magnetic phenomena and it is well-known that velocity and vortical spin of a fluid are analytically equivalent to magnetic force and electric current. Or, we may identify vortical spin in the fluid medium with magnetic force and the momentum of the fluid in the medium with electrokinetic momentum. In either case, we have to take the density of the fluid medium to be proportional to its magnetic permeability. It is impossible to say which of these more nearly represents the truth, as both fail equally to account for many of the fundamental facts and both fail to stand the test of direct experiment. For the velocity of light has, so far, been found not to be affected alike by strong magnetic and electric fields, in the direction of magnetic or electric force, as it should be, if either theory at all nearly represented the truth, unless we postulate a very high density for the ether.

329. It is found, moreover, as we have already indicated, that the equations of Maxwell which the above theories attempt to visualise cannot explain all known phenomena. Modifications of these equations proposed by Hertz fare no better. They alike fail to explain such phenomena as dispersion which obviously depend on the inter-action of matter and ether. This is hardly to be wondered at. For they proceed on the postulation of two entities, matter and ether, and it is *a priori* evident that the electrical, the magnetic and the optical phenomena are modes of manifestations of the properties of a continuous medium, the ethereal, affected by the presence of matter. But the nature of this inter-action between two such diverse entities as matter and ether naturally baffles analysis. Let us postulate certain singularities in the medium—a unit of electric charge, to be presently more accurately defined. The phenomena of electrolysis and ionisation evidence their actual existence. Experiment also leads to the conclusion that the entity we should fix our attention upon is the negative charge or 'corpuscle.' And therefore a positive charge must be related to ponderable matter in some intimate way. And, further, as a first stage in the analysis of this relation, we may make the simple supposition and one most in accordance with facts, namely that a positive charge, together with a number of units of negative charge, is the material unit, called an atom.

330. In any case, the introduction of this third entity, the electron (or minute electrified particles), formally suffices to explain all phenomena. The lines of force being associated with all electric charges, the motion of these tubes serves also to image the mode of propagation of light. These tubes are tubes of electric polarisation, along which there is electric displacement and at right angles to which there is magnetic force, that is, motion—of some kind—of the ether, which would thus seem to have the structure of 'vortex sponge,' while corpuscles would be singularities possessing inertia, from which tubes of induction diverge into the ethereal medium. A conglomeration of corpuscles would seem to give rise to singularities of a more complex structure, called atoms and molecules, but in analysing the intimate nature of these singularities, science has made but small progress.

331. Summarising, then, we may say that light is a form of energy, propagated in a medium, which for convenience may be called the ethereal medium. This energy arises from a periodic disturbance, the line of flow of energy being the direction of motion of light and the periodic disturbance, perpendicular to that direction. The energy resides in the medium as the energy of stress, so that propagation of light consists in a propagation of stress.

332. This periodic disturbance consists in the vibration of minute electrified particles and vibration of the ether at right angles to the former, which corresponds to magnetic force, and one of the displacements is rotational, while the other is translational, though it is impossible definitely to specify which is rotational and which is translational.

333. The motion of the lines of electric stress (or the lines of electric induction) gives rise to lines of magnetic stress and both move perpendicularly to themselves and perpendicularly to each other with the velocity of light. A vortical spin seems to be associated with the phenomenon of propagation, the angular velocity of this spin, apparently constituting magnetic force or electric current.

334. We have thus corpuscles or charged particles, neutral particles, positively charged particles and the ether. There is experimental evidence, on which we may base a corpuscular theory of matter, which also suggests a means of defining the relations between the first three, while on the elastic solid theory we regard only one entity, the ether with two properties, inertia and rigidity. It is conceivable that in the primordial ethereal fluid, various processes are going on to which these properties of the medium as well as its energy are due and which are also associated with the properties of gross matter, of corpuscles and of magnetic lines of force. When these intimate processes and the exact relation between them have been unravelled, we shall have a complete theory of optics.

335. In constructing such a theory we shall have to bear in mind that the various processes with which we have to deal, as well as the medium which is their seat, are conditioned by the intrinsic relativity of space and time and that these processes and the associated energy are essentially discontinuous. That, however, only modifies but does not alter the characteristic nature of the problem presented.

APPENDIX I

1. It will be useful in the first place to state my exact point of view.

We have the dynamical equation $\delta \int (T - V)\, dt = 0$ which gives a complete account of the motion of a dynamical system—in this case, the disturbed optical medium.

We have also the equation $\delta \int dt = 0$ (the dynamical significance of which requires investigation).

Are these independent of each other? If so, the second equation can only be the equation of constraint. No such constraint can, so far as we can see at present, be well associated with the medium considered.

If no such constraint can be postulated, we can only regard the second equation as identical with the first, in this particular case.

2. Now in order that we should be justified in doing so, it is necessary to admit that t has the same meaning in both Hamilton's principle and in Fermat's law.

3. It has been argued* that this is not permissible as "in Fermat's law we have $\delta \int dt = 0$ (1) or $\delta \int \frac{ds}{v} = 0$ (2), where the co-ordinates are no longer those of particles of matter as in Hamilton's principle, but successive points on the ray as the light-wave travels along, and the velocity of the ray is entirely different from the velocity of the individual material particles whose motion constitutes the light." It will be seen, however, that this argument is based on (2), which is derived from (1) by a mere analytical transformation but this would amount to imposing a limitation (from the point of view of the present line of argument) on equation (1), not necessarily involved in it. In

* Dr G. T. Walker, *Phil. Mag.* Jan. 1919.

fact, the second equation may well be taken as $\delta\int\frac{d\phi}{\phi'}$ where $\phi'=\frac{d\phi}{dt}$ and may, as such, be held to give information regarding ϕ and ϕ', *whatever these may be* (not merely s and v), so long as these quantities are related in any manner to the phenomenon of light propagation. But the particular co-ordinates involved in ϕ and ϕ' or their nature cannot well be regarded as alone implied in (1). I conceive, therefore, that this line of argument is not crucial against my theory.

4. Moreover, it stands to reason that Hamilton's principle, although it directly deals with a certain volume distribution of energy, will naturally lead to equations giving propagation of energy or disturbance, if T and V are appropriate to such a propagation. The latter equations may, in this event, involve a set of co-ordinates distinct from that involved in Hamilton's principle and yet we may not argue that these equations are independent of that principle.

5. It is now necessary to consider another difficulty, which has also been raised and which has always appeared to me of great importance. We know that Hamilton's principle postulates that the initial and final configuration of a dynamical system are prescribed and the time of transit of the system from the initial to the final configuration must remain unchanged. Now the first condition applies also to Fermat's law and the second condition *may* be imposed on it, if we take the time of transit to be that from one wave-front to the next. This, however, may deprive my conclusions of a part of their generality, but only in a manner which it is not possible to decipher at present.

6. As to the contention that "in Hamilton's principle the co-ordinates are those of particles of matter," I am doubtful whether this limitation will be universally acceptable. I rather think, given the forms of T and V, in a medium which is the seat of energy, Hamilton's principle will be applicable, though we may be unable to determine the intimate nature of the constitution of the medium which determines T and V. In order to arrive at these forms, various hypotheses have to be framed, and we thus get various forms of T and V and cor-

responding optical theories. From this point of view, the electro-magnetic theory with or without modifications may well be regarded as a dynamical theory.

7. I must admit, however, that even if the identity between Fermat's law and Hamilton's principle can be established, I have not been able as yet to *prove* that $T - V =$ constant is the *only* solution. Therefore, although I cannot think of any other solution ($T - V = f(t)$ being inadmissible, on the principle of energy), I content myself with saying (p. 12) that $\int(T - V)\,dt = 0$ will be consistent with Fermat's law *if we take* $T - V =$ constant.

8. This seems to be all the more desirable in view of what I have stated in paras. 1 and 5. I do not therefore claim to have proved that all energy is kinetic. I only suggest that Fermat's law is capable of an interpretation which will yield this conclusion. I trust the proof will be forthcoming in due course. At present, however, I have no such illusion on that point.

APPENDIX II

MAXWELL'S STRESS SYSTEM

The fundamental conception is that of Faraday, viz. that the energy in the electrostatic system resides in the field. This is at once justified as a physical interpretation of the theorem

$$\int V\nabla^2 V dxdydz = \int V\frac{\partial V}{\partial n}dS - \int\left[\left(\frac{\partial V}{\partial x}\right)^2 + \left(\frac{\partial V}{\partial y}\right)^2 + \left(\frac{\partial V}{\partial z}\right)^2\right]dxdydz$$

$$= 0,$$

or, since $\frac{\partial V}{\partial n} = -4\pi\sigma$ and electrostatic energy $= -\frac{1}{2}\int \sigma V dS$,

$$W = \frac{1}{8\pi}\int F^2 d\tau. \qquad \text{[Art. 115.]}$$

The next step is to find the stress system which will give rise to this (volume) distribution of energy.

If we assume this energy to arise from tension along a tube and pressure at right angles to it, we can find the magnitude of these.

This is tantamount to the assumption that the strain energy function of the elastic medium contains first-order terms in the strains.

Maxwell's method of procedure is different: he assumes the ether to be of the nature of an elastic medium, under extraneous forces

$$-\rho\frac{\partial V}{\partial x}, \quad -\rho\frac{\partial V}{\partial y}, \quad -\rho\frac{\partial V}{\partial z},$$

where

$$\nabla^2 V + 4\pi\rho = 0.$$

This is allowable, as Maxwell shows (*Treatise*, Chap. V, Art. 103, Vol. I), on the ordinary law of electric action $\frac{ee'}{r^2}$ without any special hypothesis as to the intimate nature of electric density in relation to the ether regarded as a material medium [Art. 120].

The solution obtained by Maxwell, however, being only a particular solution, fails to satisfy all the conditions of the problem.

APPENDIX III

Nearly all the salient points with regard to discharges in vacuum tubes of De la Rive pattern, containing air can be derived from the curves (fig. 1, I and II). In these, pressures in millimetres of mercury are ordinates, and the corresponding potential-differences between the electrodes in electrostatic units are abscissae.

Fig. 1. I and II.

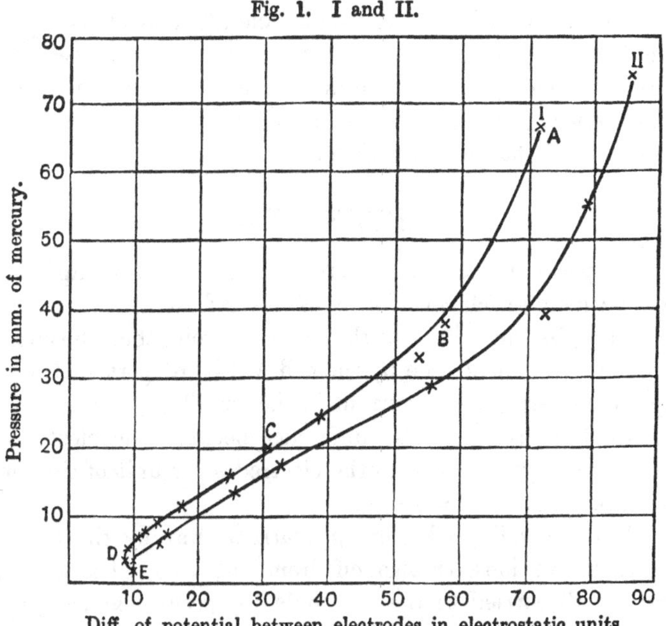

I E.M.F. of discharge due to 3 cells in the primary of the induction coil.
II. E.M.F. of discharge due to 4 cells in the primary of the induction coil.

These curves were obtained by measuring the lengths of sparks between two brass spheres of 3 cm. diameter *in parallel* with the discharge tube and deducing the corresponding potential-difference from the table given at p. 461 of J. J. Thomson's *Conduction of Electricity through Gases*, 2nd ed., by interpolation.

It will be seen that we may roughly distinguish four portions in each curve, *AB*, *BC*, *CD*, *DE*. The first portion (*AB*), which is very nearly straight, corresponds to the 'spray' discharge. As the pressure decreases the character of the discharge changes; it forms into a band by the confluence of most of the discrete streams. This is indicated by the bend in the curve at *B*.

After this, the relation between pressure and potential-difference is given by a straight line *BC*. Throughout this stage, the discharge is in the form of a band of light which rotates according to the law $p\omega =$ constant. Gradually the curve bends away from the straight line *BC*, and at this point (somewhere about *C*) it appears that the above law ceases to hold. Ultimately it bends round, as is also *a priori* evident from the fact that at a very low pressure the resistance to the passage of discharge is very great *.

APPENDIX IV

The peculiarities of the curves, fig. 1, seem to be capable of explanation on such considerations as the following:

Let V_0 be the voltage of the induction coil; then the energy supplied per unit of time by the coil will be proportional to V_0, say $i_0 V_0$, where i_0 is the current in the circuit.

Let V be the potential difference between the electrodes; then the energy supplied to the electrodes per unit of time will be proportional to $V = i'V$, say.

Therefore $i_0 V_0 = i' V +$ energy carried away by the positive and negative ions, thrown off from the electrodes, *less* the energy (E) carried to the electrodes by positive and negative ions reaching them (per unit of time).

* *Phil. Mag.* Oct. 1908. Recent experiments have shown that similar curves are obtained with all vacuum tubes.

But the energy carried off by an ion $= Xe\lambda$.

Therefore

$$i_0 V_0 = i' V + Xe(Nq'\lambda' + nq\lambda) - E,$$

where n and N are the numbers of positive and negative ions thrown off from the electrodes and occupying unit length of the discharge, and λ, λ' their mean free paths.

In order to find E, we may proceed as follows:

It can be shown * that the equations of continuity in a discharge-tube can be written, in the steady state,

$$\left.\begin{aligned} \frac{\partial Nq'}{\partial x} &= \alpha Nq' + \gamma nq \\ -\frac{\partial (nq)}{\partial x} &= \alpha Nq' + \gamma nq \end{aligned}\right\} \quad \text{......................(1)},$$

where

$$\left.\begin{aligned} \alpha &= \frac{1}{\lambda'} f(Xe\lambda' - \beta') \\ \gamma &= \frac{1}{\lambda}[F(Xe\lambda - \beta)] \end{aligned}\right\} \quad \text{..................(2)},$$

and n, N the number of positive and negative ions per unit length of discharge, x being measured along the line of discharge.

Therefore, we have

$$Nq' + nq = \text{const.} = \frac{i}{e},$$

where i is the current carried by the discharge.

Again, the energy carried to the cathode by the positive ions may be written equal to

$$\frac{eVx}{d} e^{-kx} (\alpha Nq' + \gamma nq),$$

where k is a coefficient determining the dissipation of energy during the passage of these ions.

Also, the energy carried to the anode by the negative ions may similarly be written equal to

$$\frac{eVx'}{d} e^{-k'x'} (\alpha Nq' + \gamma nq)$$

* J. J. Thomson's *Conduction of Electricity through Gases*, 2nd ed. p. 490.

where d = distance between the electrodes;

V = difference of potential between the electrodes assumed to vary uniformly.

$$\therefore \quad E = \frac{eV}{d}\int (\alpha N q' + \gamma n q)\, x\, (e^{-kx} + e^{-k'x})\, dx.$$

But from (1), if α, γ be regarded as constant,

$$(\alpha N q' + \gamma n q)(\alpha - \gamma) = \alpha \frac{\partial N q'}{\partial x} + \gamma \frac{\partial n q}{\partial x};$$

$$\therefore \quad \alpha N q' + \gamma n q = (\alpha N_1 q_1' + \gamma n_1 q_1)\, e^{(\alpha-\gamma)x} \quad \text{.........(3)},$$

if $N = N_1$, $n = n_1$ at the cathode.

Hence

$$E = \frac{Ve}{d}(\alpha N_1 q_1' + \gamma n_1 q_1)\int_0^d x e^{(\alpha-\gamma)x}(e^{-kx} + e^{-k'x})\, dx$$

$$= \frac{Ve}{d}(\alpha N_1 q_1' + \gamma n_1 q_1)\left[\frac{d e^{(\alpha-\gamma-k)d}}{\alpha-\gamma-k} - \frac{e^{(\alpha-\gamma-k)d} - 1}{(\alpha-\gamma-k)^2} + \text{similar terms in } k'\right]$$

$$= Xe(\alpha N_1 q_1' + \gamma n_1 q_1)\, P, \text{ say.}$$

If $\alpha = \gamma = 0$, the above equation reduces to $E = 0$. This we may suppose to be the case during the rotatory stage in air [*Phil. Mag.* Oct. 1912]. Therefore, since in this case $N = n$, and the pressure varies inversely as the mean free path, we get the equation

$$i_0 V_0 - i' V = neX\lambda'\left(q' + q\frac{\lambda}{\lambda'}\right).$$

But $\quad i_0 = i' + i = i' + ne(q + q');$

$$\therefore \quad V_0 - \frac{i'}{i_0} V = neX\lambda' \frac{q' + q\dfrac{\lambda}{\lambda'}}{i' + ne(q + q')}.$$

As, moreover, during this stage V is small compared with V_0, and i' should be small compared with i_0, we get the simple equation $\left(\text{putting } \lambda' = \frac{1}{p}\right)$

$$V_0 = \frac{X}{p}\left(\frac{q' + q\dfrac{\lambda}{\lambda'}}{q + q'}\right) \quad \text{or} \quad \frac{X}{p} = \text{const. nearly.}$$

This, as we have seen, is the case in air (curves I and II, fig. 1); when, however, the pressure is sufficiently reduced, α, γ are no longer zero. In fact, the terms in E become sufficiently effective in making V large, for α, γ, k, k' are all proportional to pressure, and it is reasonable to suppose $\alpha < \gamma$, since

$$\alpha \propto \frac{1}{\lambda'} \text{ and } \gamma \propto \frac{1}{\lambda}.$$

This is experimentally verified.

Although it is not possible to work out completely the theory of this variation of pressure without a knowledge of α, γ, k, k', we may get some insight into its nature in special cases by proceeding as follows:

From (3), we have

$$(\alpha N_1 q_1' + \gamma n_1 q_1)\, e^{(\alpha-\gamma)d} = (\alpha N_2 q_2' + \gamma n_2 q_2) \quad \text{......(4)},$$

if $N = N_2$, $n = n_2$ at the anode, $q = q_2$ and $q = q'/q_2'$;

but $$e\,(N_2 q_2' + n_2 q_2) = i = e\,(N_1 q_1' + n_1 q_1) \quad \text{.........(5)}.$$

If, now, $n_2 = 0$,

$$\frac{\alpha i}{\alpha - \gamma}\, e^{(\gamma-\alpha)d} - \frac{\gamma i}{\alpha - \gamma} = e N_1 q_1',$$

We have also

$$i_0 V_0 - i' V = X e N_1 q_1' [\lambda' - \alpha P], \text{ since } n_2 = 0;$$

and as $\frac{i'}{i_0}$ is a small quantity, we get X as an exponential function of p.

It is obvious, however, that the above investigation is not capable of giving a complete account of the variation of the potential difference, for we have assumed (1) that the potential varies uniformly from cathode to anode, and (2) that α, γ are constant. As neither of these suppositions can be true always, it is not surprising that the curves obtained* are more complicated than those given by theory.

* *Phil. Mag.* July, 1916.

APPENDIX V

An approximate theory of the magnetic action of the excited magnetic field on electric discharge through a De la Rivé tube may be worked out as follows:

Using cylindrical coordinates, z, ρ, θ (where z is measured from the Faraday dark space), the equations of motion of an ion may be written, if m is its mass,

$$m\ddot{z} + A\dot{z} = Ze + He\rho\dot{\theta} \quad \text{.................(1)},$$

$$m(\ddot{\rho} - \rho\dot{\theta}^2) + A\dot{\rho} = R \quad \text{.................(2)},$$

$$m\frac{1}{\rho}\frac{d}{dt}(\rho^2\dot{\theta}) + A\rho\dot{\theta} + B\theta = He\dot{z} \quad \text{...........(3)},$$

where $A =$ coeff. of viscosity, B a coeff. to be determined, $Z =$ electric force in the direction of z and R in the direction of ρ, while $H =$ magnetic force, which we know is mainly in the direction of ρ (*Phil. Mag.* Jan. 1908).

Now, considering the equation (3) (to which, alone, we shall confine ourselves), if we have N negative and n positive ions per unit length, in any stream (masses m_1 and m_2 respectively), we have, taking moment about the axis and summing

$$\int(m_1N + m_2n)\,ds\frac{d}{dt}(\rho^2\dot{\theta}) + \int(A_1N + A_2n)\,\rho^2\,ds\,\dot{\theta}$$
$$+ \int B\rho\,(N+n)\,ds\,\theta = \int H\rho e\,(N+n)\,ds\,\dot{z} \quad \text{......(4)}$$

as the equation of motion of any stream of discharge.

Now we may assume, as in *Phil. Mag.* Oct. 1912, the action between two streams of lengths ds, ds' to be a repulsion

$$= \frac{e^2 ds\,ds'}{Kr^2}\left[(N-n)^2 - \frac{(nq + Nq')^2}{3V^2}\right] + \frac{\pi\rho}{r^3}(na^2q - Nb^2q')^2,$$

where q, q' are the velocities of positive and negative ions, a, b their radii (assumed spherical), r the distance between ds, ds', K the S.I.C. and V the velocity of light.

The third term of (4) will then be of the form

$$Cf(a)(n-N)^2\theta,$$

the other terms (depending on the velocities) being neglected. Here C is a constant depending on the form of the various streams of discharge, and α the angular coordinate defining the position of the stream, whose equation of motion is given by (4), provided n and N are constant throughout the discharge, for in this case, alone, θ will be the same for all points. In any case, if $n = N$, the equation of motion is of the form

$$I\ddot{\theta} + \mu\dot{\theta} = \int H\rho i\, ds$$

= the couple acting on the discharge due to the magnetic action of the electromagnet,

where $$I = \int (m_1 + m_2)\, n\, ds\, \rho^2,$$

since $\dot{\rho} = 0$, in the steady state.

But this couple $= \frac{3}{2}Mi$, where M is the total magnetic strength of induced magnetism (*Phil. Mag.* Oct. 1908).

Therefore we have

$$I\ddot{\theta} + \mu\dot{\theta} = \tfrac{3}{2}Mi,$$

where $$\mu = (A_1 + A_2)\, n \int \rho^2 ds.$$

This is the same equation as was obtained in Art. 263 by identifying the discharge with an electric current.

If the number of positive particles is small in comparison with that of negative particles, the number of the latter will not necessarily be constant throughout any stream of discharge. In this case, putting $n = 0$ and considering the motion of a small element of a discharge, we have, when the steady stage is reached,

$$B\theta = He\dot{z} \quad \text{or} \quad \theta = \frac{He\dot{z}}{B} \quad \text{..............(5)},$$

where B is a function of ρ, α, defining the position of the element of the discharge considered. This completely explains the twist referred to in Art. 258*.

* *Phil. Mag.* July, 1916.

APPENDIX VI

QUANTUM THEORY

1. It was proved by Kirchhoff that the energy of radiation in an enclosure of which the sides are maintained at constant temperature is distributed according to wave length, so that the density of radiant energy for any interval $d\lambda$, or $E_\lambda d\lambda$

$$= f(\lambda, T)\, d\lambda,$$

where T is the absolute temperature of the enclosure, f being independent of the nature of the enclosure.

Again, according to the law discovered by Wien, which can also be proved from thermodynamic considerations,

$$E_\lambda d\lambda = \frac{A}{\lambda^5} F(\lambda T)\, d\lambda \quad \text{.................(1)},$$

where F is a function of the product λT.

2. It is not, of course, possible from thermodynamical considerations alone to determine the nature of the function F, for the obvious reason that the phenomena of radiation involve atomic and corpuscular motions.

It can be shown, however, on the basis of classical mechanics, that if the energy of a system is expressible as a sum of squares of a set of variables, representing the degrees of freedom of such a system, then the total energy is divided *equally* among all the variables.

3. Let us apply this theorem to the case of a radiating enclosure of finite volume, v. The number of free vibrations that the ethereal medium enclosed is capable of, lying between the wave lengths λ and $\lambda + d\lambda$, may be taken to be $f(\lambda)\, d\lambda$, and will be proportional to v. In fact, it has been shown by Jeans and Lorentz to be equal to $8\pi v \lambda^{-4}\, d\lambda$ (which, indeed, can but for the constant factor be deduced from the theory of dimensions).

Again, since the number of squared terms to which each vibration gives rise is two and mean energy, kinetic or potential, corresponding to each vibration from the kinetic theory of gases is $\frac{1}{2}RT$, the energy of radiation of such an enclosure, for the range $d\lambda$, is equal to $8\pi RTv\lambda^{-4}\,d\lambda$, or $8\pi RT\lambda^{-4}\,d\lambda$ per unit volume.

This is the formula obtained by Lord Rayleigh and Jeans, determining the partition of energy, according to Newtonian Mechanics. But this cannot be true, for if we integrate the expression from $\lambda=0$ to $\lambda=\infty$, the total energy becomes infinite for any finite value of T.

4. Planck has, in fact, deduced from the experiments of Lummer and Pringsheim that

$$A\lambda^{-1}F(\lambda T)=\frac{8\pi h\nu}{e^{\frac{h\nu}{RT}}-1},$$

where h is a constant, known as Planck's constant, which is equal to $6{\cdot}6\times10^{-27}$ erg sec.$^{-1}$ nearly, and ν, the frequency, corresponding to λ.

If we put $h\nu=\epsilon$, then the total energy (for the interval $d\lambda$) becomes

$$8\pi\lambda^{-4}\,d\lambda\,\frac{\epsilon}{e^{2h\epsilon}-1}.$$

In order to interpret this, we note that in the limit, when $\epsilon=0$, this reduces to $8\pi\lambda^{-4}\,d\lambda\,\frac{1}{2h}$, or Lord Rayleigh's expression, if

$$\frac{1}{2h}=RT.$$

In other words, if the principle of equipartition of energy were true, there would be nothing to distinguish one wave length from another, the total energy would not involve any quantity of the type ϵ, and there would be nothing to distinguish any one portion of space from any other.

Since, therefore, this principle is not experimentally verified, we must impose on the distribution of energy a certain discontinuity, the nature of which is indeed suggested by the form of Planck's formula.

5. Let us recall, for this purpose, the well-known theorem* that the probability that a system shall have its coordinates

* Jeans, *Dynamic Theory of Gases.*

$(p_1, p_2, \ldots)$ and momenta $(q_1, q_2, \ldots)$ within a range $dp_1, dp_2, \ldots dq_1, dq_2$, is proportional to e^{-2hE}, where $2hRT = 1$, and E is the energy in this configuration.

Now, it is reasonable to suppose that the same law of probability will apply to *all* groupings—those of vibrations for instance—in the system. Considering, then, a certain range of vibrations between λ and $d\lambda$, of which the number N is $8\pi\lambda^{-4}\,d\lambda$, let us suppose that n of these has zero energy, then $ne^{-2h\epsilon}$ may be expected to have energy ϵ, $ne^{-4h\epsilon}$ the energy 2ϵ, etc. Then, we must have,

$$N = n(1 + e^{-2h\epsilon} + \ldots) = \frac{n}{1 - e^{-2h\epsilon}},$$

and the total energy of these vibrations

$$= n\epsilon e^{-2h\epsilon} + 2n\epsilon e^{-4h\epsilon} + \ldots$$

$$= n\epsilon \frac{e^{-2h\epsilon}}{(1 - e^{-2h\epsilon})^2} = \frac{N\epsilon}{e^{2h\epsilon} - 1}$$

$$= 8\pi\lambda^{-4}\,d\lambda \frac{\epsilon}{e^{2h\epsilon} - 1}.$$

6. Comparing this result with Planck's, we are led to his theory of quanta, which states:

The energy corresponding to a degree of freedom of frequency ν cannot increase in a continuous manner but only by quanta or finite magnitudes, the amount of which is proportional to ν.

Poincaré* has given a rigorous proof of the above principle, and has in fact shown that "whatever be the law of radiation, if we suppose that the total radiation is finite, we obtain a function, presenting discontinuities, analogous to that of the hypothesis of quanta."

* Jeans, *Report on Quantum Theory.*

INDEX

Aberration, 59
— electron theory of, 115, 148
— Fresnel's theory of, 118
Aristotle, undulatory theory, 4

Bartika, commentary of *Nyaya Vasya*, 3
Boussinesq, 152

Cauchy, 55
Characteristic function, 8, 9
Corpuscles, 142
Current, 74, 83

Descartes, theory of refraction, 4, 5, 6
— theory of light, 6
Diffraction, 23, 92
Dispersion, 59, 60
— electron theory of, 148
Double refraction, 16, 24–30, 48

Elastic solid theory, 33
— modified, 55
Elasticity, theory of, 35
Electric displacement, 66
— physical meaning of, 73
Electro-kinetic energy, 85
— momentum, 84
Electro-magnetic theory, 61
Electro-magnetism, 80
Electron theory, 124
Electrons, 102
Electrostatic energy, 65, 130

Faraday, 71
Fermat, 6
Fresnel, 21, 25, 118

Glazebrook, 55, 58, 96, 115
Green, 45, 51

Hall's phenomena, 114
Helmholtz, 153
Hertz, 88, 121, 124
Hindus, 2
Hooke, 13
Huyghens, 21, 23
Hydrodynamic analogy, 107

Induction, magnetic, 78
— normal, 63
Ions, 142

Karl Pearson, 1
Kelvin, theory of discharge of a Leyden jar, 88
— labile ether, 40
— vortex sponge, 161
Ketteler, 154
Kinetic, all energy is, 12
Kirchhoff, 94

Labile ether, theory of, 40
Larmor, 48, 101, 103, 106, 110, 133, 137
Lodge, 110
Lorentz, 104, 138

MacCullagh, 28, 43, 51
Magnetic energy, 80
— permeability, 79
— shell, 81
Magnetism, 76, 129
Magneto-optic energy, 114
Magneto-optic rotation, 109

Malus' law, 30
Mean free path, 143
Michelson-Morley experiment, 160
Molecular vortex theory, 111

Newton, 13–20
Nyaya Kandali, 3
— *Sutra*, 3
— *Vasya*, 2

Optical energy, 91

Poincaré, 70, 132
Poisson, 38
Poisson's solution, 93
Polarisation, 24
— rotatory, 59
Polarising angle, 31
Potential electrostatic, 62
— magnetic, 76, 81
Poynting, 91
Pythagoras, 4

Quantum theory, 161

Raman, 97
Rayleigh, 54, 55, 93
Reflection and Refraction, 14, 31, 46, 98
— metallic, 99
Relativity, Theory of, 163
Resonator, 89
Röntgen, 122
Rowland, 95

Saint-Venant, 68
Specific inductive capacity, 62
Stokes, 39, 43, 95
Strain-energy function, 38
Stresses in the electrostatic field, 67
Swiftest propagation of light, 6

Thomson, Sir J. J., 89, 100, 117, 132, 134, 141
Tubes of force, 63
— moving, 119
— Faraday, 134

Undulatory theory, 4, 20

Vaisesik School of Hindu Philosophy, 3
Vector-potential of magnetic induction, 84
Vedanta Parivasa, 3
Verdet, 114
Vibrator, 88
Visual influence, 4
Vortex-sponge theory, 161

Young, 21

Printed in the United States
By Bookmasters